ICE

Two young women deliver ice during World War I (1918).

ICE

FROM MIXED DRINKS *to* SKATING RINKS—A COOL HISTORY *of a* HOT COMMODITY

AMY BRADY

G. P. PUTNAM'S SONS · NEW YORK

PUTNAM
— EST. 1838 —

G. P. PUTNAM'S SONS
Publishers Since 1838
An imprint of Penguin Random House LLC
penguinrandomhouse.com

Hardcover ISBN: 9780593422199

Printed in the United States of America
1   3   5   7   9   10   8   6   4   2

Book design by Elke Sigal

For Alan

CONTENTS

Introduction · XI

PART 1
THE BIRTH OF AN OBSESSION · 1

Chapter 1
The Man Who Would Be Ice King · 3

Chapter 2
Blasphemous Ice · 31

Chapter 3
The Iceman Cometh · 51

Chapter 4
Highways, Holidays, and the Cheap-Ice Revolution · 71

PART 2
FOOD AND DRINK · 87

Chapter 5
The Inside Scoop · 89

Chapter 6
"Culture in a Glass" · 111

Chapter 7
On the Rocks · 123

CONTENTS

Chapter 8

Fire and Ice · 141

PART 3
ICE SPORTS · 159

Chapter 9

Fancy Figures · 161

Chapter 10

Cork-Bungs, Brooms, and Zambonis · 181

Chapter 11

The Need for Speed · 199

PART 4
THE FUTURE OF ICE · 211

Chapter 12

Fevers, Freezers, and Frankenstein · 213

Chapter 13

TV Dinners, Henry David Thoreau,
and a Swiftly Warming Planet · 233

Chapter 14

Flammable Ice · 251

Conclusion · 261

Acknowledgments · 267

Notes · 269

Photo Credits · 299

Index · 301

ICE

INTRODUCTION

Harold "Red" Grange, All-American halfback,
delivers ice during the off-season (1930).

In America, ice is everywhere, in gas stations, restaurants, hospitals, our homes. Americans think nothing of dropping a few cubes in a tall glass of tea on a hot summer day. They fill large coolers with ice for weeklong camping trips. They wrap ice in dish towels and press it to the skin of their bruised arms. They skate, race, and play hockey and curling matches throughout the year on indoor ice rinks. Parties end when the ice runs out.

A 2020 poll conducted by Bosch reveals that 51 percent of Americans self-identify as "ice obsessed," consuming up to 400 pounds of ice each per year. Over 8 million refrigerators are sold in the United States each year, most of which have built-in ice makers. Yes, ice in America is everywhere, but it wasn't always this way. The nation's ice obsession is barely two hundred years old. And until now, the story of how that obsession came to be has never been fully told. I know because I went looking for it.

In the summer of 2018, a record heat wave gripped the planet, choking wildlife and straining energy grids. It hit when I was visiting family in my hometown of Topeka, Kansas. A series of rolling blackouts knocked out the air conditioner in my parents' home, so we all piled into their Chrysler and headed to a nearby gas station, which was operating on a generator. Inside, a blast of cold air felt like bliss on my overheated skin. I headed for the soda dispenser and filled a cup with ice. As I watched the cubes fall, I was struck by an irony. I knew after a decade of reading climate-science studies that heat waves like this one were consequences of climate change. The irony lay in the fact that I was driven to seek out and consume ice because of a phenomenon that's eliminating ice on the planet.

For days, I mulled over this paradoxical connection between ice and climate, wondering how ice became so prominent in America in the first place—and what its future might look like in an era of accelerating climate change. Much to my amazement, I couldn't find one comprehensive cultural history on the substance written for popular audiences. Every book I found discussed ice in terms of its scientific properties or as it relates to polar exploration. All of it was fascinating, but not what I was looking for. I realized I had stumbled into a bona fide hole in history, and I wanted to fill it like that cup of gas-station ice. For the next several years, I visited archives across the country, interviewed experts, and read everything I could find about ice. At long last, a fuller history started to present itself, and it was stranger and more deeply entwined with other areas of American life than I ever could have imagined.

———————

America's obsession with ice was sparked by climatic change, and I don't just mean the global warming scenario we find ourselves in now. The ice industry developed at the turn of the nineteenth century, which coincided with the tail end of the Little Ice Age. Unlike human-made climate change, the Little Ice Age was a naturally occurring phenomenon, a four-century stretch of planetary cooling caused by a combination of erupting volcanoes and a period of calm on the surface of the sun. It was a time of spectacularly harsh weather that brought frosts in summer and weeks of thick snow and below-freezing temperatures in winter. Lakes and rivers froze to depths of several inches or more, allowing teams of men to chisel out blocks of ice, which their bosses then sold across the country, and eventually around the world.

If you watched the 2013 Disney film *Frozen*, you've seen an animated version of these men. It opens with six burly harvesters sawing blocks of ice out of a lake and hauling them to the back of two horse-pulled wagons. The ice saws depicted in the movie remarkably resemble those wielded by actual harvesters. But real-life ice harvesting wasn't as pleasant as the film suggests. Catchy songs and cute animals trading wry looks aside, real-life ice harvesting was dangerous business. Lakes and rivers didn't freeze evenly, and the thinner spots weren't easy to identify just by looking. Men and horses frequently fell through the ice, sometimes to their deaths. The men lucky enough to be rescued were brought to the ice shanty, where they were handed a cup of whiskey to drink and warm themselves up. The horses wore lengths of rope around their necks so that if they fell through, the men could cinch the ropes tight, choking the animals just enough to keep them from struggling as they were pulled to safety.

Ice harvesting was also dirty business. Natural ice was filthy with sediment and pollution, especially during the peak of the Industrial Revolution. Horses, being the living mammals that they are, let loose their bowels on the ice—the very ice that would soon touch the mouths of thirsty Americans. Every ice-harvesting team employed a man to clean the ice by pulling a "shine sleigh" over areas where horses had just been. The sleigh scraped away the waste as he walked, thus earning its name from the shine it left behind.

As the Little Ice Age came to an end, winters grew warmer. In the United States, lakes and rivers south of Maine rarely froze to the depths required to harvest the amount of ice that Americans had come to depend on. So ice harvesting went north, until those bodies of water also failed to produce enough ice. This environmental change coincided with technological innovations and social and economic shifts that ultimately brought the American natural ice-harvesting industry to an end. But that didn't eliminate the nation's taste for ice. How could it? The people were obsessed. Instead, the American innovative spirit that gave the world the cotton gin, the telephone, and 3D printing also spawned new ways of making and distributing ice. Those innovations made ice available to more people than ever before, and by the end of the nineteenth century, it had become a resource on par with coal in importance.

Nowhere else in the world was there an ice industry as large as the one in the United States. Thomas Jefferson built one of the largest icehouses in the country at Virginia's Monticello. In a letter to his business manager, Jefferson wrote that the man must have ice wagons ready to "come at a moment's warning" lest the icehouse ever go empty. When Charles Dickens came for a visit from his home in southern England, he gawked at the American "icehouses [filled] to the very throat" and "the mounds of ices" that Americans ate in hot weather. Ice was so prominent that it was even used on the dead. The body of President Ulysses Grant was kept in a casket filled with ice for preservation in the summer heat as it was carried by train, then horse and carriage, from his cottage near Saratoga Springs, New York, to New York City, where he was eventually laid to rest.

To understand just how transformative ice was to American life, it helps to understand first how miserable life was before ice was everywhere. Before the launch of the ice industry and way before electric refrigeration, people living in warm climates—such as America's southern states and territories—had no effective way of getting cool. Temperatures didn't fall low enough for ice to form there, even in winter. So residents drank their water tepid, even when sick with fever. They preserved food as best they could by salting or canning it—two methods vulnerable to human error, so food poisoning was common, especially among children. They tried to create cold compresses for injuries using plants and pastes and other cooling substances, but the concoctions weren't cold enough to be of much help. Cocktails, including an early version of the mint julep, were served at room temperature. Imagine.

Ice was much more common in colder areas, of course, but because it was dangerous to harvest and expensive to store, only the wealthy had regular access to it. Ice was a marker of status. Those with means used ice year-round. Those without, didn't.

Not that Americans were the only ones to use ice. Far from it. The Chinese were cooling their drinks with ice at least as early as A.D. 960. The Persians were building icehouses, or *yakhchāls*, and harnessing the laws of thermodynamics to make ice under the desert sun as early as 400 B.C. The British built their first icehouses in 1660. And long before colonizing settlers came to what we now call North America, Indigenous communities relied on ice for travel, storage, and shelter. The Inuit built homes out of ice. The Iñupiat have for centuries built ice cellars to store their food. Ice is so critical to the lives and culture of the Yupik people of western Alaska that they have dozens of words for sea ice. (As the planet warms, however, some of those words are becoming obsolete.)

So why the focus on American ice, specifically?

This narrower geographic focus allowed me to tell a more complete story in a relatively small amount of space. Ice has been essential for millennia for all kinds of people around the world, each with their own cultures, histories, and uses of the substance. To tell a comprehensive world history of ice would require many more pages than could fit in a single book.

A focus on American ice also highlights the unique way in which the American ice industry began and proliferated throughout the nation. Americans became obsessed with ice not because of breakthroughs in technology or scientific understanding (though those things helped). They became obsessed because of an outrageously persuasive marketing plan.

Throughout the history of ice in America, marketing and the people who were really, really good at it emerge as revolutionary forces, changing forever how the public thought of ice and where it fit into their lives. Every time the ice industry was disrupted, whether by climate, technological innovation, or something else, it was marketing that convinced people that ice was still relevant to their everyday existence.

Marketing with this kind of reach and at this level of effectiveness isn't unique to the ice industry. History is full of moments when Americans were taught how to form their habits or fall in love with new products. The automobile industry launched one of the biggest marketing campaigns of all time to encourage people to switch from horses and buggies to horseless carriages. Eighty years after the introduction of the Model T, the computer company Apple used marketing to turn a love for its products into a lifestyle. After the $30 million shoe company Nike launched its "Just Do It" campaign in the 1980s, it grew to be worth billions. All companies and industries, big and small, rely on marketing to convince potential customers that their product is worth purchasing. The American ice industry just did it first.

The history of ice in America is told here in four parts. The first part presents the birth of an obsession, telling the story of what sparked America's love for ice and how the ice industry evolved over time, from natural-ice harvesting to the creation of manufactured ice. This section reveals how ice gave rise to the brewery and fishing industries; how it changed where and when Americans play winter sports; how it created great wealth for some people and caused widespread death among others; and how it sparked the careers of some of the most notable figures in American entertainment and politics. This first section encompasses its own narrative arc, but it tells just a fraction of ice's story.

The remaining three parts delve deeper into areas of history in which ice played an outsize role. Part 2 focuses on food and drink. It opens with a look at the history of ice cream in America. Suffice it to say no one would have had ice cream without ice—it's right there in the name. But Americans also owe their love of ice cream—and its widespread availability—to the contributions of presidents, immigrants, and Black entrepreneurs, all of whom saw in ice the power to transform tastes—as well as society. Their stories are told here. This section also explores the history of iced tea, whose origins are far more peculiar than most people realize, and the history of cocktails, which played an early role in getting America hooked on ice. Later, they would help Americans reimagine what ice looks and tastes like in the twenty-first century. The section concludes with a history of ice sculpting, a practice with roots in Indigenous cultures and the restaurant business.

Part 3 delves into sports history, looking at how America's obsession with ice nurtured a love for winter sports, specifically ice skating, figure skating, hockey, and speed skating. This section also looks at curling, because there are few sports played on a stranger sheet of ice. Readers will learn more about how the invention of mechanical ice boosted the popularity of winter sports, because it made skating possible year-round and anywhere in the country, whether in

Arizona's desert or Florida's tropical tip. Readers will also learn about the weird science of ice's slipperiness and how a team of ice experts in Utah have harnessed the unique properties of ice to create the fastest rink in the world—ice so slippery that more world records for speed skating have been broken in Utah than anywhere else on earth.

Part 4 turns to the future, looking at how ice is transforming not only America but the world as we know it. Ice fundamentally altered how medicine was practiced, saving (and, at times, taking) lives as doctors learned how to exploit the power of cold to heal bodies. Today, ice plays a role in experimental treatments for cancer that might one day revolutionize how the disease is treated in some of the world's most underserved communities. This section also looks at how ice gave birth to electric refrigeration and how refrigerators, while vital to American life, are taking a toll on the planet. The section ends with a glance at how a strange substance called "flammable ice" might just help wean the world off fossil fuels.

Taken together, these four parts reveal how ice has profoundly shaped the nation's history and culture, not only in terms of how Americans behave but also in terms of how they think. As Heidi Julavits wrote in the *New Yorker*, "The American need for ice speaks to our obsession with refrigeration as an antidote to death." If we follow her line of thinking (and indeed, this book does), then to be without ice is to allow rot to take hold. But to have an abundance of ice is to preserve that which is most precious, including our food, our bodies, our cultures, our histories. America's obsession with ice is an obsession with life.

PART 1

THE BIRTH OF AN OBSESSION

CHAPTER 1

The Man Who Would Be Ice King

Two men harvesting ice in Michigan (1903)

The history of America's obsession with ice is much like the history of America itself: steeped in myth, given shape by acts of defiance, powered by commercial interests, and driven by bold yet deeply flawed men with visions that would change the world. It could be said to begin in Boston on September 4, 1783, just a day after the Revolutionary War ended, when Delia Jarvis Tudor gave birth to her third son, Frederic Tudor. No one could have known it then, but Frederic would one day spark a revolution of another kind: a revolution in how Americans think of and use ice.

The sun was shining the day I visited the site where Tudor's childhood home once stood. A couple of blocks over, I could see Boston's Old State House, where his father, a high-ranking judge named William Tudor Sr., would have hobnobbed with merchants and met senators for midday drinks. Not far to the north is the Bell in Hand Tavern, opened in 1795 and purportedly the nation's oldest bar, whose cocktails today, clinking with ice, are an unacknowledged tribute to Tudor's influence. In 1806, Tudor would have walked a half mile east from here to the harbor, where he'd set sail on the first American ship to carry blocks of ice on the open sea. Forty years later, he would stand in that same spot and, spitting curses, toss his wife's belongings into the water.

I went to Boston to better understand this man who was crowned the "Ice King" by the city's newspapers for his success in launching the American ice trade. He was a man, I would learn, of many contradictions: a fortune seeker who frequently spent more than he made; a charming salesman with a violent temper; a shortsighted businessman but also a visionary. Where his Bostonian peers were skeptical of a plan to make money off something as common and, well, *free* as Massachusetts lake ice, Frederic wrote to potential investors asking for the capital to harvest what only he saw as a bounty to then sell in warm climates around the world. Almost everyone, except a few savvy financiers, thought he was mad for even suggesting the idea. No one had ever attempted to ship ice long distances. How would he keep it from melting? And who would buy it—and why?

Frederic, however, had a vision as well as a business plan. He also suffered a personal loss that haunted him for years, a loss that likely helped spur his desire to create a nation of people utterly obsessed with ice—and eager to pay for it.

I n February 1801, Frederic Tudor was just seventeen years old and had long been a thorn in his father's side. William Sr. believed that success meant a respectable career in business or law, but Frederic had refused to go to college—not even to Harvard, where the judge's business partners would have granted him admission. Instead, Frederic insisted on trying one moneymaking scheme after another, learning the ropes by experience—and mostly failure—rather than by the books. Exasperated, his father asked Frederic to join his brother, John Henry, who was suffering from bouts of weakness and a knee condition, on a months-long trip to Cuba. The trip, he figured, might force Frederic to stop his scheming and rethink his career path. Or so he hoped.

Frederic agreed—who would turn down an all-expense trip to the tropics?—and later that month, the brothers boarded a ship named *Patty* in Boston Harbor. They carried suitcases full of wool suits—the fashion of the day, even in warm temperatures—and nearly two hundred boiled eggs. The family chef had told them eggs would travel well. They did not.

As the ship departed, the brothers' minds were likely filled with dreams of sun-drenched Havana, an image that would have been in stark contrast to the cold, gray harbor stretched out before them. This had been an especially stormy winter in New England, and the sky roiled with ominous clouds. The ship set sail anyway, and within hours, a downpour flooded the deck, forcing the men to their cabins. The incessant rocking of the ship soured their stomachs. They spent the next several days sitting on their cots, bent at their waists, heaving into bedpans.

The ocean grew calmer once they reached southern waters. The sun came out, burning their noses and zapping the seawater collecting on the deck, turning it to steam. The brothers were at a loss for how to cope with the heat. Some of the crew suggested they avoid the sun by staying in their cabins, but without good ventilation, the tiny rooms were intolerably hot. The trip lasted four weeks.

When the *Patty* finally docked in Havana, the brothers death-marched onto shore with, as John Henry later wrote, "their tongues hanging out." Surrounding them were palm trees swaying in a light breeze. A cerulean blue ocean lapped at white sands. Paradise, it seemed, was everywhere they looked. Frederic hired an English-speaking servant named Francisco, who at first was something like a godsend: he introduced the men to pineapple juice and rum, fresh oranges, sweet guavas, and green figs. But when he noticed John Henry's poor health, Francisco (along with the woman who owned the house the brothers were staying in) restricted his diet to rice and water, an intervention that did nothing except annoy John Henry, who was determined to eat whatever he wanted, even if he had "to fight Francisco, kick the old woman & tell the doctor he's a quack." John Henry's knee grew worse.

Spring turned into summer, which brought even warmer weather, which in turn brought clouds of mosquitoes and armies of scorpions. The insect invasion was followed by a raging epidemic of yellow fever, a disease that ravaged the Caribbean and much of the American South every year. Doctors and scientists had yet to develop a vaccine for the disease, largely because they didn't understand that mosquitoes transmitted it through their stinging bites.

People throughout Havana fell ill with "the shakes"—a symptom of high fever that could last for days. When both Tudors caught this, they didn't know how to treat their aches and pains. Back home, they would have asked a servant to chisel chunks of ice from a block stored in the family icehouse so that they could press the cool substance to their hot foreheads. But warm places like Cuba didn't have icehouses, or even frozen water. The temperatures never dropped low enough, and a slap to the ocean's tepid surface would send a spray to the face that felt like sweat.

When other Americans left, so did the brothers. They purchased passage on a ship to Charleston, South Carolina, the first ship back to the United States they could get on short notice. Their luck didn't improve. The ship turned out to be carrying molasses in its cargo, which gave off a stench that oscillated between overripe fruit and rotting meat. Meanwhile, the heat was unbearable, and there was no way to get cool.

Eventually they made it to Philadelphia, where they parted ways. Frederic went back to Boston, and John Henry stayed behind. John Henry's health continued to fail, and in January 1802 he died with his mother by his bedside. His death shook Frederic to his core. For the next year, Frederic lived in alternating states of rage and regret. If only he had had access to ice in Cuba, he thought, perhaps his brother would have survived. Today we know that ice wouldn't have saved John Henry. He likely suffered from bone tuberculosis, a rare disease that requires specialized treatment developed only in the last one hundred years. But Frederic didn't know that. As far as he knew, John Henry died because he couldn't get cool.

Over the next several months, Frederic's behavior grew erratic. His family and friends avoided him, so he bought a diary to record his rants. Death and revenge on his mind, he drew a tombstone on the cover. Inside the drawing he wrote the following: "He who gives back at the first repulse and without striking the second blow, despairs of success, has never been, is not, and never will be, a hero in love, war or business." This vague threat aimed at no one in particular would become something of a motto for Frederic for the rest of his life.

———

During the nineteenth century, most wealthy families living in cool climates like New England enjoyed ice year-round, thanks to icehouses. These "houses" were more like stone cellars built on private land that went several feet into the ground where temperatures, even in summer, rarely rose above fifty degrees. George Washington owned a sizable icehouse at Mount Vernon in Virginia, which his enslaved servants famously stocked to make a novel dessert called ice cream. In 1793, Martha Washington wrote that "in the warm season Ice is the most agreable [sic] thing we can have."

Frederic Tudor knew the comforts of coolness. As one of the wealthiest families in Boston, the Tudors owned a large icehouse that their servants kept filled with blocks of ice year-round. Every winter, the Tudor servants risked their lives by cutting 100-pound blocks of ice out of Rockwood Pond, a lake not far from the Tudors' country estate. They'd stand several yards out on the slick, frozen water, where the frigid wind blew hardest but the ice grew deepest. They carved the ice by hand using large saws, praying with every stroke that the frozen water held their weight. Once the ice broke free, the men used long metal rods to float the blocks as close to shore as possible. There, another half dozen men with horses waited to heave the blocks onto land and into the back of a wagon. Back at the icehouse, servants carried the blocks several feet to the bottom, where they stacked them vertically then padded them with sawdust and straw to keep them from melting. If packed tightly enough together to prevent airflow between them, the blocks would last the whole year.

The Tudor family used ice in many ways. Frederic liked to drop chunks of it into glasses of wine. The Tudor men preserved their freshly hunted game between blocks. When anyone fell ill or suffered an injury, servants nursed them back to health by pressing ice chips wrapped in cloth to their aching bodies.

These were likely the uses Frederic had in mind when, on a night in August 1805, he sat at his father's mahogany desk to sketch out the beginning of a grand plan. What if, he wondered, he could ship ice to warm climates? Ice was practically nonexistent there, ships could arrive within a few weeks, and who could say how much people would be willing to pay for coolness, comfort, and health?

But Frederic feared competitors. "Some enterprising Yankee," he wrote in his diary, might get wind of his idea and "be induced to attempt the thing" before he could try it himself. He needed to start with somewhere farther away, a place no one else would attempt to trade ice, a place such as Cuba, whose climate he believed was responsible for his brother's death.

He consulted a merchant friend of his father's on the legality of his plans. The friend explained that Frederic would have to secure permissions from the country's local governments before he could sell ice there, and Cuba made those permissions almost impossible to get. Disappointed, Frederic set his eyes instead on Martinique, a French-colonized island in the Caribbean. "Our plan now is . . . to sail by the first of November for Martinique," he wrote in his diary, "and at St. Pierre solicit the French government for this exclusive privilege." Initially, Frederic had every confidence that things would work out. "There can be very little doubt . . . they will be so pleased with the idea that the grant may be readily obtained," he wrote.

The scheme's greatest challenge, as far as Frederic was concerned, would be securing the capital to harvest enough ice to make the trip worthwhile. In the following weeks, he asked every well-to-do connection he had for a loan, including Harrison Gray Otis, a former associate of his father's and by then a state senator. Frederic wrote the man a letter saying he had it on "good authority" that ice can be preserved on ships for long stretches of time at sea. (He actually had no evidence of this.) He referred to regular shipments of ice from Norway to London as proof that his plan would work. (There's nothing in the historical record to suggest that a Norwegian ice trade had yet been established.)

The senator refused, writing that the plan wasn't for him but still "sounds plausible." Frederic confined his annoyance to his diary, where he wrote, "Whether [the senator] did think it plausible or no, he is too much of a politician for *me* to find out." In a letter to another potential investor, he dropped a line that would make Arnold Schwarzenegger's ice-pun-making villain Mr. Freeze from the 1997 *Batman & Robin* movie quite proud: "For heaven's sake, do not be too *cold* to the thing and miss a passing opportunity." (Emphasis added.)

While he awaited response, another problem was beginning to dawn on him. Frederic couldn't be in Boston overseeing the ice harvesting and in the tropics getting trade permissions at the same time, so he turned to his older, twenty-six-year-old brother, William, for help. William, whose heavily lidded eyes were often shielded by a mop of unkempt hair, wasn't thrilled with the idea of going into business. He was a poet who eventually cofounded the *North American Review*, not an entrepreneur. But once he learned that the job included luxury travel (or so he thought) to the tropics, he agreed to assist. For company on the voyage, William recruited their cousin James Savage. Frederic agreed to the choice because the cousin had "some business habits and character in him," welcome contrasts to what he saw as William's perpetual mooniness.

On a warmer-than-usual November day, Frederic, now twenty-two, stood on the dock of Boston Harbor and waved as his brother and cousin set sail on a ship called *Jane* for Martinique. The decision to start their venture there may have seemed to make business sense, but it proved a dangerous choice. This was 1805, and a war had recently broken out between Britain and France. When the *Jane* was more than halfway to Martinique, it came face-to-face with a British warship that insisted on inspecting what the vessel was carrying. This wasn't unusual. As a neutral nation during the Napoleonic Wars, the young republic engaged in trade with both warring countries—an agreement that had made many American traders of the era very rich. But this also meant that American ships were often subject to search— and on occasion, to questionably legal seizure. Eventually the British soldiers were satisfied with *Jane*'s civilian status and let the crew go. *Jane* sailed on.

The night they approached Martinique, the sea was tranquil. In an era before electricity, the Milky Way would have stretched above them in a twinkling ribbon of light. Then, suddenly, came cannon fire. The shot fell just short of the bow. A frantic crew member climbed the shrouds and hung a lantern to signal *Jane*'s neutrality. The firing stopped, but everyone aboard agreed to drop anchor for the night just in case the conflict resumed. The next morning, with all still clear, the ship sailed into the port of Saint-Pierre.

I n the late eighteenth century, as revolution shook France and its colonies, the French legislature formally abolished slavery. But this was no help for the enslaved of Martinique, as the British seized the island from the French in 1794, never allowing the abolition to take root. By the time William and James arrived in 1805, Saint-Pierre was home to more than thirty thousand enslaved people and their owners—all of whom produced waste. The waste was carried away by streams of gray water that gushed through the cobbled streets of town. A putrid smell filled the air from dawn to dusk, and the heat, even in November, somehow thickened it. Somewhere along the journey, both William and James contracted a virus—possibly yellow fever—that left them bedridden for days. When they were finally well enough to schedule meetings with local authority figures about trade permissions, no one would see them. They thought the idea to sell ice was too strange.

Still weak, the men half walked, half limped from office to office until, finally, a local prefect granted them the permission they sought, after accepting a bribe of two gold coins. The contract allowed the men to start selling ice immediately. Satisfied, they set sail again, this time for Jamaica.

This stretch of the trip went almost as poorly as the last. They were on a different ship, and this ship's captain was deep in his cups by ten each morning, rarely leaving his cabin after the first drink. This left the duties of steering the ship to the first mate, who was largely unfamiliar with the ship's mechanics and nearly capsized the vessel twice. Everyone agreed to dock in Santo Domingo to search for a new captain to lead them on. It was an unplanned stop, but since they were there anyway, William and James figured they'd apply for permission to sell ice. The local government granted their request, but that was the end of their good luck. French soldiers occupying the Dominican Republic prevented them from traveling to any other island. They managed to find passage on a Danish schooner bound for Jamaica, but a privateer stopped them just outside Port Royal and took William's favorite pistols. When they were finally permitted to proceed to port, they stayed for several weeks just to avoid having to get back on a ship. "A seaman's life is a dog's life," James later wrote in his diary.

The men eventually sailed back to Boston, getting lost along the way for nearly twenty-seven days. When they finally arrived, a flustered and annoyed Frederic met them at the harbor, showing little sympathy for their bad luck. To his mind, they'd spent way too much money and had accomplished little. "The advantages derived from their part of the expedition," he wrote in his diary, "were not equal to the expense of it, which was near $2,000." When he calmed down, he explained how busy he'd been while they were gone, ginning up interest among investors (who, in actuality, had either said no outright or pledged to talk again after Frederic proved his plan could be successful) and mortgaging the Tudor family's South Boston land to cover the rest of the capital. This would be the first of many poor business decisions.

He'd also been working on a plan to keep ice cold while at sea. Frederic had noticed that ice in the family icehouse melted fastest in open air or when sitting in its own meltwater—the runoff from melting ice. He drew up blueprints for a cargo hold that would keep the ice elevated and as sealed off from air as possible. He'd even bought his own ship, the *Favorite*. This would be the second of many poor business decisions. Traders rarely owned their own vessels in Frederic's day. Instead, they purchased space on cargo ships that sailed their preferred trade routes. The ship alone cost $4,750 ($120,000 today), and reconstructing the cargo hold cost more.

William and James responded to all this by vowing never to sail again (though they did). Still, celebration was in order. As far as the men were concerned, the hardest parts of the plan were complete. They had permissions to sell ice. They had their ship. Plans were underway to begin harvesting Rockwood Pond as soon as temperatures were cold enough to freeze the lake. What could go wrong?

The next challenge wouldn't be the considerable labor of extracting and shipping ice. What none of them had considered was how they would actually sell the ice, a substance people in the tropics had rarely—if ever—seen.

Even as their plan came together, no one seemed to have any faith in what the men were doing—not even Judge Tudor, who told his son that no one in their right mind would pay money to buy ice when they could get it for free. (The judge rarely traveled outside Massachusetts.)

"People only laugh and belittle me when I tell them I am going to carry ice to the West Indies," wrote Frederic in his diary. "Let those laugh who win."

As the shipment day grew closer, Frederic had more to worry about than ridicule from family and friends. There existed the real possibility that he wouldn't get the ice harvested in time. This had been one of the warmest winters on record, with ice thinner than it had ever been in Frederic's lifetime. Not until two weeks before the February shipment date did Rockwood Pond freeze deep enough to harvest thick blocks that could survive, at least in theory, the weeks-long journey to Martinique.

Frederic hired dozens of men—more than the Tudors had ever employed at one time—to carve up the lake and haul the ice blocks to the harbor, where they were hoisted into the *Favorite*'s newly constructed cargo hold. When the ship was full, Frederic strode aboard and walked to the bow. As the ship nosed its way toward the open sea, he would have looked out at a cold, hazy horizon. Whatever came next, he was about to be the first person on earth to bring ice to the tropics. And, he thought to himself, he was about to make a fortune.

Meanwhile, the press back in Massachusetts was having a field day. "No JOKE," reported the *Boston Gazette*. "A Vessel has cleared at the Custom House for Martinique with a cargo of ice. We hope this will not prove a *slippery* speculation." (Emphasis added.)

On March 5, 1806, the *Favorite* arrived at the island. Much to Frederic's delight, the carefully constructed cargo hold had prevented two-thirds of the ice from melting. The bad news was that there was no place on the island to store it. Only now did Frederic appreciate just how difficult this trading scheme was going to be. To sell ice in the Caribbean in 1806 was somewhat akin to selling television sets to Americans before the 1950s: Few broadcasting stations existed in the early twentieth century because few people owned TVs. But few people owned TVs because there were so few stations. Worse, TVs didn't become puddles if they languished unsold. To succeed at selling ice, Frederic was going to have to build both interest and an infrastructure to support it.

As he pondered this, the hot sun was melting the ice on the ship fast. Frederic had no other option but to sell it directly from the cargo hold, a move that fell into a legal gray area, because while selling goods from a ship wasn't technically illegal, only those sold on land were actually legal. He charged sixteen and a half cents per pound and sold fifty dollars' worth of ice in two days, before sales dried up. Frederic sat in his sweltering cabin, puzzling over why so few people were buying ice, when one of the ship's crewmen alerted him to an angry customer on the dock. Frederic wiped his sweaty brow, straightened the cuffs of his wool jacket, and met the man outside.

The islander gestured at Frederic and then at the ice with anger. *"Il fond!"* he shouted. It melts! Bewildered, Frederic started to explain that, yes, of course ice melts, when a second customer appeared, this one as angry as the first. The second man explained that he'd put his ice in a tub of water to stave off the melting, but the water made it melt faster. Frederic stood dumbfounded. He realized that for all his planning, he hadn't accounted for a simple fact: for the majority of people living in the tropics, a block of ice would have been as fanciful as a unicorn. No one knew how to keep ice, let alone how to use it. They wouldn't know how to carve chunks to drop in drinks, or how to crush it for making ice cream. They wouldn't have known that ice soothed swollen limbs or cooled fevers.

To educate the public, he scribbled instructions on how to carry the ice (wrap it in a blanket) and store it (keep it wrapped in a dark room). Clearly, icehouses needed to be built, and fast. He used what was left of his financial reserves to hire carpenters and stonemasons. Toward the end of the month, with the ice melting, he agreed to accept an offer of $3,000 from a merchant on the island to take a cargo of sugar back to Boston.

Once home, he shut himself in the family house to consider what to do next. His debts were due, and he had nothing to show for his work. He sought new investors, but the only interested person was another cousin, James's older brother. The brother had connections in Havana—the kind that might help Frederic get the permissions he needed to trade there. Frederic accepted his money and sent William and James to Cuba to try to convince the government to let them sell ice. Miraculously, they were successful, and that spring, Frederic sent two ships of ice to Havana. The capital city reacted similarly to the people of Martinique and the Dominican Republic. No one knew what to do with ice, and he sold hardly any of it.

Over the next two years, Frederic sent three more ice shipments to Havana, each one a dismal failure. Meanwhile, his father made a series of bad investments in Boston, losing much of the family fortune. Just when Frederic thought things couldn't possibly get worse, he was arrested for debt on State Street in full view of the business owners and their customers. The judge and his friends were able to scrape together enough bail money to keep Frederic out of prison, but Frederic was mortified, writing to his brother-in-law that the whole ordeal had been "abominable."

By 1809, Frederic was twenty-six years old and owed nearly $40,000 to investors. Three years later, the War of 1812 was raging and he still owed money to a number of people in Boston. On one of his rare walks through town, he was again arrested for debt, and this time thrown in prison. Humiliated, he sat behind bars until a close family friend bailed him out.

Back home, the few servants the Tudor family could still afford to pay attempted to cheer him with glasses of iced wine and bowls of homemade ice cream. He closed his eyes and reveled in the coolness of the drink and dessert. Something occurred to him. People unfamiliar with ice wouldn't just figure out what to do with it quickly enough for him to turn a profit. No, he had to show them—actually demonstrate in real time—how ice could be used to make the most delicious things. He wasn't just selling frozen water; he was selling a lifestyle. He knew then in his gut that once people got a taste for such treats, they would buy more ice than he had lake.

———————

It took Frederic a while to gather the funds to travel, but by November 1815, he was sailing to Cuba once more. There was a warrant out for his arrest that would double his jail time if he were to leave the state. To get around this, Frederic asked his ice harvesters to load the cargo ship under the cover of night, which they agreed to do at twice the pay. (Ice harvesting was especially dangerous in the dark.)

The day he arrived in Havana, he went looking for local cafés. If there was anything his previous times in Cuba had taught him, it was that no one on the island trusted him—but they did trust their local baristas. Café culture was dominant in Cuba. Each barista had their own twist on café Cubano or a proprietary recipe for mixing crushed fruit with rum. These beverages, as delicious as they were, were served lukewarm. Tudor's innovation was to change that. He offered several pounds of ice to dozens of competitive baristas for free, on the condition that they'd allow him to demonstrate how best to serve their drinks chilled. And because the ice cost the baristas nothing, they could charge the same amount for chilled drinks as lukewarm ones. Though skeptical, some agreed.

At first, customers were suspicious of the chunks of frozen water floating in their cups. But you can't argue with the pleasure of a drink on the rocks. Word of the chilled beverages spread, and demand grew quickly. When the cafés ran out of ice, the baristas asked Tudor for more, which he sold, and at an ever-steepening price. Part one of his plan was complete—he had created a market and was finally turning a profit. Now it was time for part two.

Frederic returned to the baristas with a second proposal: he'd provide more ice for free if they'd allow him to demonstrate how to mix it with salt, fresh cream, and fruit to make a delicious frozen treat. *El helado*, or ice cream, became another sensation, and demand for ice spiked once again. For the first time in nearly seven years, Frederic sold all his ice. When he returned to Boston, Frederic made plans to ship more ice to Havana within a year.

Demand in Cuba kept growing, and iced coffee and cocktails were becoming popular all over the country. Ice cream, however, had become a veritable national obsession. The frozen dessert eventually became so popular among Cubans that by the end of the nineteenth century, the country was importing ice cream from the United States to augment its own production and keep up with demand. The revolutionary leader Fidel Castro was famously smitten with ice cream. When the United States implemented a Cuban trade embargo in the 1960s, he asked his Canadian ambassador to send him twenty-eight containers of ice cream—one of each flavor from Howard Johnson's— just so he wouldn't run out. So large was Castro's appetite for *el helado* that the novelist Gabriel García Márquez once recalled watching him consume eighteen scoops, even after "a good-sized lunch."

The dessert was equally popular with the general public. As tensions between Cuba and the United States worsened, *"helado por el pueblo"* (ice cream for the people) became a rallying cry among revolutionaries. Castro took up the slogan as part of his political platform and, with cash from allied Russia, built the largest ice-cream parlor in the world. He called it the Grand Cathedral of Socialist Ice Cream. Its enormous dining area sat up to one thousand people, and its menu boasted twenty-six flavors. Alongside ice-cream classics like vanilla and chocolate were Cuban originals such as pineapple glaze and coconut almond, each made with local nuts and fruits.

Today, Castro's parlor still sells scoops for just five cents each—a price that's still within the average Cuban's personal budget. Whereas Frederic Tudor had seen ice cream as the key to capitalistic success, Fidel Castro saw it as vital to creating a socialist utopia.

B ack in Boston, Frederic reveled in his triumph. Money from the ice trade rolled in. The newspapers, in a series of mea culpa articles, crowned him the "Ice King" in their headlines. Frederic had conquered the tropics; now it was time to bring business to the southern United States. His fear of competitors was what finally convinced him to make his move. Having proved that shipping ice was possible, he figured it was only a matter of time before someone else tried doing the same closer to home.

Frederic set his eye on New Orleans, a major city crammed onto a tiny slip of land hemmed between the Mississippi River, two lakes, and an enormous swamp. At only 110 miles upriver from the Gulf of Mexico, it was one of the busiest trading ports in the northern hemisphere. It was also blisteringly hot in summer—the perfect place to sell ice.

Frederic turned again to William for help, but by then his brother was experiencing some minor literary success with a book called *Letters on the Eastern States*. He turned down the offer. Frederic next turned to his youngest brother, Harry, the black sheep of the family. Grateful to be taken seriously by his brother, Harry eagerly agreed.

Frederic sent Harry to New Orleans to oversee the building of icehouses near the port. They planned to send their first shipment by the end of the year. All was going well until Frederic received a note from Harry saying that one of the icehouse keepers had gotten drunk and bragged that Frederic was bringing ice to the city. Still concerned about copycat competitors, Frederic dashed off an angry letter. "It is absolutely necessary that I should be obeyed," he wrote. "I must be a dictator."

In the winter of 1821, a ship full of Frederic's ice set sail for New Orleans. (Frederic decided to stay behind to figure out what to do about his sagging international ice sales.) As the ship neared the port, its crew would have seen dozens of merchant ships and flatboats lined up by the docks, seeking to trade cotton, coffee, and sugar.

Harry must have learned from Frederic how to spark interest in ice, because Frederic praised him heartily in his diary for his sales-manship. One can imagine Harry waiting for the ship to arrive, gathering ice from its cargo in a bucket, and then heading out to find the city's busiest drinking holes. He would have found them in the French Quarter, the heart of New Orleans, which was in the midst of a major cultural transition from residential neighborhood to bustling market-place. Multistory town houses made of wood lined the dirt-packed Bourbon Street, their top floors rented to low-income residents while their first floors functioned as saloons. The sounds of pianos playing African songs and the melodies of European composers filled the streets during the day, before brass bands took over at night—the city's innovative musicians who, drawing on the many cultures that came together in the Mississippi Delta, would eventually give birth to jazz. The clientele at these places was a mix of working-class locals and wealthy traders who came to blow off steam.

The first saloons that Harry visited would have resembled the Old Absinthe House, which still operates today at the intersection of Bourbon and Bienville Streets. Patrons could choose to sip their drinks in main barrooms, which flooded with light from the floor-to-ceiling windows, or outside in courtyards framed by tall picket fences covered in ivy.

The bartenders were just as impressive. Drawing on Creole and Spanish recipes passed down through generations, they mixed their cocktails with the precision of scientists. With Frederic's mentorship, Harry would have known that in order to convince them to add ice to their meticulously crafted drinks, he'd need to do more than demonstrate how to drop ice in a glass. He would need to wow them with spectacle.

Harry might have described his cocktail recipe as something akin to magic—it literally changed the temperature of the liquor. Hearing this, bartenders would have invited him behind their bars, where Harry could make a show of surveying his choice of ingredients. The Tudors' go-to recipe called for sugar and mint, which Harry would have crushed together in the bottom of a glass. In a taller glass, he would have poured gin over a scoop of irregular chunks of ice from his bucket, covered the glass with a towel, and given it a shake. When the liquid cooled, he would have poured it over the mint. This was a smash, similar to a mojito, a family favorite. Perhaps Harry gave the drink one last stir for showmanship before handing it to the bartender, whose eyes must have gone wide when the cold liquid touched their lips.

Just as Frederic had done in Havana, Harry offered the bartenders of New Orleans several pounds of free ice to test whether customers would like it. And just as in Havana, customers liked it very much. Harry must have demonstrated his mixology talents to any bartender who would let him, because he sold what was left on the ship in days. When Frederic got word of this back in Boston, he was confident that Harry had turned New Orleans into an "ice city." He had in actuality done much more than that: Harry and the ice he sold had changed the city—and cocktail culture—forever.

As ice became more available, the city's bartenders experimented with its shape and size to learn how it altered the taste and consistency of their creations. They found that ice could be chipped from a block or cracked to a size that would approximate today's ice cube. It could be crushed with a small hammer into sparkling slivers that made cocktails twinkle in the afternoon sunlight as if full of diamonds. Bartenders soon discovered that different forms of ice changed the flavor and balance of their drinks. Shaved ice, for example, diluted cocktails quickly—this added a new element of mixology to experiment with. If dilution was not desired, they used large cubes or simply swirled a chunk of ice around the lip of the glass to cool it before filling it.

Within the next couple of decades, bold New Orleans mixologists invented some of America's most famous cocktails, including the rye-based Sazerac, considered by many tour guides (and locals) to be the city's signature drink. Today, New Orleans is known across the country and around the globe as "the cradle of drinking culture."

J ust as Frederic had predicted, copycat competitors flocked to New Orleans. By 1840, fleets of ships were carrying ice south along the eastern seaboard, while railroads, which had only just begun to crisscross the country, brought ice to the Midwest and West Coast. Frederic boosted business by expanding into other southern ports and eventually to international destinations, including India. When international trade didn't turn as much profit as he'd hoped, he ventured into ice-tool development. He worked with a former ice harvester turned inventor named Nathaniel Jarvis Wyeth to develop and market a horse-drawn ice cutter that sped up ice harvesting and created more evenly cut blocks.

As ice became more available, its price fell, enabling more Americans to afford it at least a couple of months a year, when it was most plentiful. They stored their ice in new contraptions called iceboxes, invented in 1802 by a dairy farmer named Thomas Moore. These wooden boxes were approximately four to five feet tall and lined with tin. The ice fit into a top cabinet with a hole in the bottom. As air flowed over the ice and through the hole, it cooled the air in the lower cabinet, keeping whatever was stored there chilled for hours. This development transformed America's relationship with food. Most Americans prior to the ice trade salted their meat to keep it from spoiling, but that was an imperfect method. Fruits and vegetables tended to rot quickly, especially in summer. The icebox lowered rates of food poisoning, improving America's overall health. "Ice must be considered as outdoing most other luxuries," Frederic wrote in his diary. In the eyes of many Americans, it was actually becoming a necessity.

I f, as Alexis de Tocqueville once argued, America's number one defining characteristic is its pursuit of independence, a close second might be its pursuit of beer. In 2021, the American beer industry raked in $100 billion in sales. The industry wouldn't have existed without Frederic Tudor's ice trade.

In the early nineteenth century, the western shores of Lake Michigan were home to a substantial German population, who brought their family recipes for lager beer when they immigrated to America. Made from a fermentation process that starts at the bottom of a barrel and works its way upward, Midwestern lagers were light and crisp on the tongue, much smoother than the ales of New England. Lager can be brewed only at cold temperatures, so few people outside the region had tasted it. That's because, while ice was available to Midwesterners living near the Great Lakes, there hadn't been an infrastructure to harvest it in great quantities. With the arrival of Frederic's horse-pulled ice cutters, commercial-sized icehouses, and a growing workforce to run them, local family brewers expanded their businesses and began to bottle lager for shipping by train. In 1844, the Pabst Brewing Company opened in Milwaukee, followed by the Miller Brewing Company, and then Anheuser-Busch in St. Louis. As demand for ice in the upper Midwest grew, wannabe ice kings flocked to the region, where they were as territorial and competitive as the ones out east.

By 1900, the ice and brewery industries were well established, and two Milwaukee-based ice companies were about to prove that no business was as cutthroat as the ice business. For years, the Great Lakes region had been served exclusively by the Wisconsin Lakes Ice and Cartage Company, also known as Wisconsin Lakes—a misnomer, because it cut ice from the Milwaukee River. That changed when the Pike and North Lake Company, or Pike Co., moved in and set up shop on Lake Michigan. This was during the peak of the Industrial Revolution, when the Milwaukee River—like most rivers that ran through cities—was burbling with chemicals and other pollutants. Pike Co. launched an advertising campaign that touted its lake ice as the purest in the Midwest, while subtly hinting that any company still harvesting from rivers was passé. This incensed Wisconsin Lakes, which responded with an act of revenge that rivaled anything Frederic Tudor might have come up with.

The plan was ruthless but not rash. Wisconsin Lakes examined Pike Co.'s infrastructure for weaknesses and found a big one. Pike Co., it turned out, had built its icehouses conveniently close to the railroad but had failed to purchase the land between the houses and the tracks. In a clandestine visit to the county commission, men from Wisconsin Lakes obtained the deed to those intermediate lands. They waited until Pike Co. workers had harvested thousands of pounds of ice and were just about to transport it to the railroad, then slapped them with a no-trespassing order.

This, Pike Co. decided, was war.

Like Wisconsin Lakes, Pike Co. took its time to properly research its revenge. It reviewed the types of commercial permits available in winter and saw that winter-season steamship river rides were not only legal but also quite popular among tourists. It purchased an old but sturdy steamship and outfitted it with an icebreaker. The company kept it docked downriver near a touristy riverwalk, where no one from Wisconsin Lakes would ever suspect it belonged to Pike Co. The day that Wisconsin Lakes ice harvesters began cutting, Pike Co. distributed a flyer advertising inexpensive "river excursions" featuring "live music." Tickets sold out. As the ship departed, a brass band played ragtime favorites on the deck, their sound carrying south on the wind and right into the ears of the Wisconsin Lake ice harvesters. The band kept playing as the ship smashed through the ice, sending the men running, and shattering the ice into millions of unsalvageable splinters. The scene was like *Titanic* in reverse.

This went on for weeks, until a slick ice king from California swooped in, bought both companies, and consolidated them. Overnight, the rivals became colleagues and were forced to work side by side.

Frederic Tudor died a wealthy man in 1864 at the age of eighty, leaving behind a wife and six children. Not much is known about Frederic's romantic life, except that he chased women and few chased him. His first long-term courtship began in 1829 with Elizabeth Hopkinson, whom Frederic described as a "charming twenty-nine-year-old widow" from Philadelphia and Bordentown. The forty-seven-year-old bachelor met her at a ball in Nahant, Massachusetts, where her family had brought her to ease her back into society after a year of grieving for her late husband.

Frederic courted her for the duration of her stay in the city. After she left, he sent dozens of letters, but he lacked his brother's talent for poetry. Instead of compliments or vows of devotion, he wooed her with the current selling price of ice. Two weeks before Christmas, Frederic learned from a mutual friend that Elizabeth was ending the affair, and on December 23, he received a package containing every letter he'd ever sent her. If he felt spurned, he never let on. In his diary he wrote, "Free life, no marriage."

That sentiment didn't last forever. Four years later, he met Euphemia, the woman he'd marry when he was fifty-two and she was nineteen. They held a small ceremony after a brief courtship, and shortly thereafter, Euphemia was pregnant. Sadly, she miscarried. She blamed her "delicate troubles" on her husband's sexual appetite, which she found irritatingly insatiable. In a matter of weeks, she was pregnant again, and her health deteriorated. When Euphemia started to show symptoms of a second miscarriage, Frederic called on two doctors to help her. The first claimed she was still pregnant, the second concluded she was not. The second doctor was right.

Over the next seventeen years, Euphemia would miscarry at least once more but eventually give birth to six children. Frederic, who'd always found personal relationships difficult, rarely wrote about their marriage or children, but others did. The poet Henry Wadsworth Longfellow recounted a dinner party with the Tudors, during which Frederic "flamed like a comet from the top of the table," insisting that a "wife is the natural enemy of the husband." Another story tells of an argument that ended with Frederic throwing Euphemia's belongings into the harbor. The only explanation for this outburst is in a letter that Frederic wrote to his wife, to which she unsurprisingly never replied: "I am 64 years old & you are not yet 34. . . . It was necessary to harden you a little in the affairs of life, in case my vigor & spirit should leave me too soon."

Euphemia told more of her side of the story in the most salacious place she could—Frederic's own diary. After his death, she annotated the book with her interpretation of events. Among them was a charge that Frederic had cheated on her. Among the most likely candidates were the owner of the boardinghouse where they first met and their housekeeper of many years. Frederic never fully admitted to anything in his diary, but he confessed to having a wandering eye. He ended the self-damning passage with "my children must excuse as best they may." Euphemia felt otherwise. "Truth," she wrote in the margins, "is a strong antagonist!"

Frederic's interaction with his children remains mostly a mystery, though his diary suggests they were not close. Perhaps the most damning evidence that they had a strained relationship lies in the choices of their eldest son, Frederic Jr. Unlike his father, junior attended Harvard, where he studied engineering. Upon graduating, this son of the Ice King became a pioneer in household heating.

CHAPTER 2

Blasphemous Ice

Ice harvesting on Conneaut Lake, Pennsylvania (1907)

B y the time the Civil War ended in 1865, nearly 750,000 American soldiers had been killed in battle or died from disease or starvation. During the war's second and bloodiest year, President Lincoln signed the Emancipation Proclamation, ending slavery in the States and changing the Old South forever. No war since the Revolutionary War altered the political, social, economic, and cultural landscape of the United States as much as this one did.

One year before the war ended, Frederic "the Ice King" Tudor stared out his window and watched the ice harvesters pull blocks from the lake on his family estate. The blocks wouldn't be traveling far. A wartime embargo had brought the southern ice trade to a halt, causing the region's chefs, bartenders, nurses, and doctors to lose access to the northern ice they'd come to rely on for preserving food, making drinks, and healing bodies.

Without ice, southern hospitals in particular were suffering. In the mid-nineteenth century, American medicine still resembled what was practiced in the Middle Ages. Germ theory was not yet widely accepted among medical professionals, and chloroform, ether, and whiskey were standard-use anesthetics—when an anesthetic was used at all. When soldiers were wounded in battle, many preferred to lie bleeding in the barracks over going to a hospital, where they risked infections from other patients. Still, when most medical treatments failed, ice sometimes helped reduce fevers and swelling, and sometimes helped stave off contagion. At the very least, ice made patients more comfortable. Without it, soldiers were miserable—and dying. "Sick soldiers in Augusta [Georgia] were perishing for the want of ice," the Confederate officer Jabez Lamar Monroe Curry told a newspaper in 1862. The loss of life was tragic, but equally tragic is the fact that much of this loss could have been prevented.

What Civil War–era doctors and nurses didn't know—what almost no one, save Frederic Tudor and his associates, understood—was that twenty years before the war began, a relatively unknown doctor living in the port town of Apalachicola, Florida, had found a way to bring ice to the South that didn't depend on shipments from the North. Against all conventional scientific—and religious—thinking of the day, the doctor had discovered how to make ice himself.

The doctor was John Gorrie, and his ice-making machine would eventually change, like Tudor's natural ice trade, how Americans use and think of ice. It made ice possible during ice famines—those winters that weren't cold enough to effectively freeze lakes and rivers—and even in hot summers down south after northern shipments ran out. Manufactured ice eventually made indoor skating rinks possible, changing where and when winter sports were played. A direct line can be drawn between Gorrie's invention of mechanical ice and air-conditioning, modern refrigeration, and state-of-the-art medical treatments such as therapeutic hypothermia and cryosurgery, which use ice crystals to freeze tumors.

None of that mattered to Gorrie, though, who died a laughingstock at the relatively young age of fifty-one. He considered himself a failure and blamed his poor fortune on Frederic Tudor. In some ways, he was probably right.

On a warm spring day in March, I visited the John Gorrie Museum in Apalachicola, a one-room building more Florida roadside attraction than proper museum. On the way, I stopped at a nearby gas station to buy a large soda. As I filled a cup with ice, an attendant behind the counter asked if I knew who'd made that ice machine possible. He looked surprised when I told him I did.

"No one knows about Gorrie," he said as he rang me up. "He's a legend around here. They named the bridge after him." That would be the John Gorrie Memorial Bridge that carries US 98 and US 319 over the Apalachicola Bay. It was built in 1988, replacing an earlier model from 1935. Besides the museum, it's the only significant public memorial dedicated to the doctor.

I explained that I was hoping to write Gorrie's story, and he grew serious. "You can't write about Gorrie without writing about yellow fever," he said. "It used to kill so many people, and he wanted to do something about it."

Death, it seems, has played a recurring role in the history of ice. It motivated Frederic Tudor to launch the ice trade, and it inspired John Gorrie to create an ice machine. But the parallels between the men stop there. Whereas Tudor possessed the audacious personality of a top salesman, Gorrie was an introvert who kept to himself. He also lacked the wealth needed to fund his experiments, a problem exacerbated by his inability to connect well with others. Potential investors were rarely interested.

In light of this, the fact that Gorrie achieved even a working prototype of his ice-making machine is a kind of miracle. And in fact, that's how his machine was first received—as a terrifying, blasphemous miracle.

Gorrie's childhood and teen years aren't well documented. The John Gorrie Museum claims he was born on October 3, 1803, in Charleston, South Carolina, but other sources say he was born on the

Caribbean island of Nevis. Wherever he came from, records show that when he turned twenty, he took an apprentice position at an apothecary in Charleston run by Dr. Samuel Green, one of just a handful of doctors in the city and by far the most well-known by locals. (To earn money on the side, Green also served as the local postmaster.) Working alongside the doctor, Gorrie acquired a rudimentary understanding of anatomy and common treatments for ailments such as headaches and rashes.

Throughout most of the year, their patients suffered from ordinary illnesses like colds and minor aches and pains. Then came "fever season," a stretch of four or five months when yellow fever returned to the South. Nineteenth-century doctors understood little about the disease, except that it reappeared every year like clockwork. They called it "yellow fever" because jaundice is a common symptom. Patients sick with the virus experience high fever, convulsions, savage headaches, and leg pain. Four days after symptoms start, patients either feel better or are dead. A turn toward the worst is usually indicated by bleeding in the stomach, which, when vomited, resembles coffee grounds. Hence, yellow fever was sometimes called "the black vomit."

Yellow fever killed up to 50 percent of those with severe infections, and between 1790 and 1860 it killed more than two hundred citizens every year in Charleston alone. So great was the fear of the disease that its victims were buried immediately, without visitation periods or wakes. White gauze hung on front doors, lampposts, carriage houses, and gates to signal that an infected person was on the property. During fever season's peak weeks in late July and early August, Charleston appeared to be covered in giant spiderwebs.

Out of fear, people kept their distance from each other when out in public. Grocery stores moved half their produce stands outdoors to create more space for distancing, and post offices built large wooden crates by their front doors so that mail could be dropped off outside.

Some people covered their noses and mouths with cloth in hope of warding off illness, though there was little consensus among the public whether face coverings did any good.

For nearly ten years, Gorrie worked with Green to administer experimental treatments to ease patients' symptoms. They mixed boiled pine tar with the roots of a flowering plant called mullein for patients to boil and drink. Another solution contained mercury, chloride, and a heaping spoonful of dried powder made from a Mexican root plant called jalap. It smelled terrible, and it left patients feeling dizzy and nauseated.

Relief of any kind was hard to bring to feverish patients in the subtropical South in the early 1800s. Bodily temperatures couldn't be lowered artificially, because ice was rare there in winter and all but nonexistent in summer. The ice trade wouldn't arrive for another twenty years. Even a splash of water was difficult to procure for people living in densely populated cities like Charleston, because indoor faucets hadn't been invented yet. Water had to be pumped outside, and it was always warm. Indoor toilets were also several years off, so the healthy and the sick were forced to share chamber pots. Rich families might have ventured outdoors to private outhouses, which, while at least ventilated, still reeked of waste and attracted wild animals.

After one especially deadly summer, Green convinced Gorrie to apply to medical school. The following spring, Gorrie was accepted into the College of Physicians and Surgeons of the Western District of New York on a scholarship, and that fall, he traveled by coach nearly a thousand miles north to start classes. There, he would have studied physics, biology, and chemistry, along with the most advanced medicine of the day, which by current standards was not very advanced at all. Many doctors still prescribed bloodletting and purging to balance the humors. When he graduated med school three years later, Gorrie pondered where to set up his practice. Despite haunting memories of patients with yellow fever—or perhaps because of them—he chose to go to where some of the worst outbreaks occurred in North America: the Florida Territory.

In February 1833, at the age of twenty-nine, Gorrie left for the tiny Gulf Coast town of Apalachicola. When he arrived, the weather was still relatively cool, and Gorrie spent his days sitting outside on his office porch, smoking a pipe and reading medical textbooks. At first, the locals didn't know what to make of the new doctor in town. In their view, he was a Charleston-raised sophisticate who had been educated in New York. Why had he chosen to set up practice there? Florida wasn't yet a full-fledged American state, and Apalachicola was little more than swampy wilderness. Street brawls between white settlers were common. So were violent clashes between settlers and members of the Seminole tribes, who, by authorization of the Indian Removal Act, were being forced west into Oklahoma.

Then spring turned to summer, bringing a wave of yellow fever that wiped out 69 percent of those who caught it in coastal Florida. The sick arrived at Gorrie's door jaundiced and dehydrated and shivering with fever. He soon had so many patients that he converted the entire second floor of his home into a sick ward. Gorrie treated them with concoctions mixed just as Green had taught him. When those failed, he applied leeches taken from nearby swamps to his patients' aching joints with the hope that their bloodsucking would do the sickly good. Few people got better. By August, the beds on the second floor were full, and the doctor found himself turning people away.

The outbreaks lasted longer in Apalachicola than in South Carolina because the town was surrounded by swamps on three sides. And swamps breed mosquitoes, which carry the disease. But doctors had yet to discover this. They blamed the outbreaks on swamp gas, a ghostly phenomenon that reoccurred every year at roughly the same time as yellow fever. The gas was produced overnight by vegetation rotting in the marsh. In the early morning hours, when the gas was at its thickest and smelliest, local authorities cleared the air by shooting cannons into the water. The thunderous shots rattled household windows, awakening babies and sending dogs barking, but of course did nothing to slow the spread of disease.

Gorrie, like many physicians of the day, considered another summertime phenomenon as a culprit: a mysterious vapor the American Medical Association called "crowd poison." This seemed especially insidious to doctors because, unlike swamp gas, crowd poison wasn't visible to the naked eye.

Another key difference: crowd poison didn't exist.

As the historian Salvatore Basile reminds us, nineteenth-century physicians believed that crowd poison—the idea that the breath exhaled from large numbers of people can grow toxic—explained not only the physical distress but also the godawful smell that people experienced in crowded and overheated buildings. Theaters and places of worship were often packed with people, and inside temperatures could rise to one hundred degrees during the summer. Stages and pulpits were lit with open flames, further contributing to the heat, and the buildings' mostly ornamental windows often didn't open. A theater critic once described sitting through a summer play as "a two hours seething with four thousand mortal men and women in a huge caldron of brick and mortar."

People frequently fainted in such stuffy, overheated conditions—hence the term "poison"—but the heat was rarely blamed as the cause. In the early 1800s, Americans still embraced an old European attitude toward temperature. It was cold—not heat—that was regarded with deadly seriousness. Snowstorms were thought to carry disease, and even the slightest threat of "catching a chill" sent people racing for medical help. Heat, meanwhile, was considered at worst a nuisance, something to be taken in stride.

For years, Gorrie linked yellow fever to crowd poison, because both appeared during the summer months. But eventually he began to consider that the relationship might be of correlation rather than causation. Both came with the heat, and both passed with it, too. Why not apply the same principle to a patient? That is, why not try to lower their body temperature?

There was one glaring reason for why not: the only way to lower a patient's body temperature was with ice, and in 1840s Florida, ice was almost impossible to come by. Florida had entered the ice trade just eight years earlier, and Tudor was still building his empire. Ice was dubbed "white gold" by locals because only the wealthiest Floridians could afford it, let alone owned icehouses, and Apalachicola was not a prosperous town. When shipments arrived from the North in winter, ice cost as much as five dollars per pound (nearly thirty dollars today), and in summer it was nonexistent.

While today we expect that the position of doctor might bring with it comfort and prestige, Gorrie was not a rich man. Like his mentor, Green, he had been moonlighting as the postmaster in the winter and spring—his least busy months—to earn money on the side. He finally saved up enough to buy a small amount of ice and the raw materials needed to store it. He finished his icebox just in time. The next month brought Apalachicola's worst yellow fever outbreak in history.

During the summer of 1841, yellow fever took the lives of nearly 25 percent of Florida's population, including that of the territory's former governor Robert Raymond Reid. It also killed his twenty-two-year-old daughter, Rosalie, and a week later, Rosalie's three-year-old daughter. As the illness swept through town, Gorrie put his ice theory into practice, experimenting with different ways to apply ice to patients' bodies. He found that rubbing it directly on a patient risked injuring their skin. Wrapping the ice in cloth left the patient damp and prone to additional illness, or so went the thinking of the day. He turned his attention to the room around him. If he couldn't lower his patients' temperatures directly, perhaps he could reduce the heat of the surrounding air?

Drawing on his college courses in thermodynamics, wherein he learned that cool air fell and hot air rose, he sketched a contraption that he hoped would get physics to work in his favor. He drilled a hole into a clean metal bedpan and filled it with ice, then hung the pan from the ceiling. He opened a window to encourage a breeze, and as the air flowed over the ice and through the hole, it cooled before swirling downward. Within minutes, the room began to feel cooler. Gorrie had created one of the world's first air conditioners.

The cool air seemed to ease his patients' suffering, even if they didn't immediately—or ever—get well. But this glimmer of hope was enough to convince Gorrie to keep suspended bedpans full of ice over each of his patients' beds. He went through his small ice reserve quickly, while the epidemic raged on. Many patients died. He would have seen through his office window the funeral director's men carrying the bodies out to a horse-drawn wagon, which brought them to the cemetery for immediate burial. It was during this summer of death that Gorrie had an epiphany. Even if he had all the money in the world, he still wouldn't be able to buy enough ice to keep every patient cool, because ice simply wasn't that prevalent in Florida. If he wanted more ice, he would have to make it himself.

The next three years of Gorrie's life were marked by obsession. He worked long days, dividing his time between caring for patients and experimenting with ice making. Somehow, he maintained a romantic social life through all of this, courting an Apalachicola local named Caroline Frances Myrick Beman. They married after six months of courtship.

There's no evidence that either of them was religious, but it's hard not to wonder what Beman thought of Gorrie's research. At the time, and especially in small towns, ice was considered to be God's creation—not the province of man. If word got out that Gorrie was playing God, trying to freeze water himself, his reputation would be ruined.

For the moment, that dilemma mattered little, because his experiments were not going well. The ice-making prototype he built wasn't freezing water. But it *was* leaking. It took a mechanical accident to turn his luck around. While experimenting with air pressure to drain a bucket, he discovered that the quick compression and expansion of air had a cooling effect. When he applied the compressor to water, it formed a hair-thin sheet of ice crystals across the top. The blasphemous doctor had finally done it. He'd created ice.

The machine had to be cranked by hand, and its output was slow, but it was capable of creating a lot of ice. Gorrie had ideas for increasing the machine's speed—his patents reveal he considered adding a water pump, a steam engine, even a horse to power his machine—but such improvements would cost money that he didn't have.

In the weeks that followed, Gorrie revealed his discovery in a letter to Dr. Alvan W. Chapman, another physician who had recently moved to town. Gorrie saw Chapman as a man who embraced science and modern medicine, and therefore as a potential ally, someone who would be excited—instead of horrified—by his invention. His reaction was less enthusiastic than Gorrie had hoped. His curt response to Gorrie's claim "I have made ice" was "The hell you have!"

In the spring of 1844, Gorrie began writing a series of articles in Apalachicola's newspaper, the *Commercial Advertiser*, entitled "On the Prevention of Malarial Diseases." These were his first public admissions that he'd begun experimenting with ice making. Gorrie knew what an explosive claim it was, especially in religious, small-town Florida. Turning water into ice? Why, that was impious. *Ungodly.* Worried for his reputation and possibly his safety, he adopted "Jenner" as a nom de plume in tribute to Edward Jenner, the discoverer of the smallpox vaccine.

When the first article ran on April 6, 1844, public reaction ranged from indifference to disgust. The editors of *Scientific American* published a letter reading, "We do not know of any feasible plan for producing ice artificially." Such derision made its way around the world, where the *Bombay Times and Journal of Commerce* called the invention a "cock-and-bull story." Hiding behind his fake name, Gorrie continued to write, publishing eleven articles in all. In his final one, dated June 15, 1844, he tried to convince the public of the benefits of ice making by outlining its economic impacts. He explained that Florida's ice trade relied heavily on an itinerant workforce. Even Apalachicola, which was small compared with other southern port towns, was made up of a largely transient population. Ice workers would arrive in winter and depart in the spring, carrying their profits north. Gorrie called this cost-effective structure "the most destructive system of absenteeism that ever impoverished a country." If he could build ice machines at the commercial scale, he argued, he could create permanent jobs throughout the South. The article did little to convince the pious.

Not long after, someone revealed Gorrie's identity as the author of these articles. It's unclear who did the unmasking, but it may have been an editor who sympathized with his readers' concerns that

Gorrie's machine would damn them all. Gorrie knew he needed to find potential investors quickly so that he could improve his machine and prove his critics wrong. With the last of his savings, he traveled to New Orleans, where Frederic Tudor's men had introduced ice approximately twenty-three years earlier.

G orrie almost certainly didn't understand the obstacles that Harry and Frederic Tudor had faced there in 1821, let alone the tactics they'd used to overcome them. The Tudors had relied on novelty and spectacle to sell ice, the wow factor of tasting cool liquid on demand. Even if Gorrie had understood this, the world was now a different place. The early 1840s were times of shock and change in America: President William Henry Harrison died after just one month in office. The world's first telegram was sent by the U.S. Capitol. Across the ocean, Queen Victoria was planning to marry Prince Albert in one of the biggest wedding celebrations the Western world had ever seen. And in New York City, a circus entertainer named Phineas T. Barnum had just introduced a peculiarly small boy named Charles Stratton, a.k.a. General Tom Thumb.

To capture the attention of the public, a machine that created ice—a substance that New Orleanians were still trying to figure out how to incorporate into their daily lives—needed a hook even more thrilling than the one Tudor had come up with. Gorrie didn't have it. He approached investors with facts and figures, the science behind his ice-making invention, its importance to health care, and how it fit into various economic models. They declined their support, and Gorrie returned to Florida empty handed.

Out of money, Gorrie explored opportunities in town to fund his experiments—opportunities he knew would be few and far between. He found, however, a willing publicity partner in Florida's French consul, Monsieur Rosan, who, unlike most Apalachicolians, embraced a more continental view of the world. Rosan loved spectacle, and he loved the idea of a machine that could create ice.

Every summer, the consul hosted a Bastille Day party at his home. It was an event covered by every newspaper from St. Augustine to Tallahassee. Gorrie figured if he could demonstrate what his machine could do at the party, in front of guests who'd possibly be open to the idea of human-made ice, he might just spark the interest of an investor or two. The party was held on July 14, 1850. The day was sweltering, and local stores of ice from up north had long run out. Women in heavy, ankle-length dresses fanned themselves on the divan, while men in dinner jackets mopped their brows with handkerchiefs. Someone lamented that even the wine tasted too warm. Rosan rose to his feet and tapped his glass with a knife. Everyone turned to face him.

"On Bastille Day," he announced, "France gave her citizens what they wanted. And so Rosan will give his guests what they want—cool wines! Even"—he gestured toward the parlor doors—"if it demands a miracle!"

The doors swung open, and waiters strode through, carrying small blocks of ice on silver trays. Sandwiched between the blocks were cold bottles of champagne. The guests gasped and applauded. Someone asked: Where had the ice come from? Massachusetts? Maine? The consul shook his head and held out his glass for a waiter to fill it. The ice, he told them, had been created right there in Florida.

As Gorrie had predicted, Florida's biggest newspapers ran the story. But what Gorrie didn't know was that the consul's party guests included close associates of Frederic Tudor, a man who did not take threats to his empire lightly. And for the king of the ice trade, human-made ice was certainly a threat.

Outside the party itself, this miracle was not greeted warmly. Damning headlines streaked across newspapers up and down the Eastern Seaboard, soiling Gorrie's name. Read one editorial in the *New-York Daily Globe*: "There is a crank down in Apalachicola, Florida, a Dr. John Gorrie, who claims he can make ice as good as God Almighty." Gorrie was astonished by what some people today might call a "hit piece," especially from a New York City paper. A conversation with the consul revealed that some of his guests were Tudor's colleagues, and Gorrie knew—just knew—that the Ice King was behind the smear campaign, even if he couldn't prove it.

———————

The next five years remain a hole in Gorrie's story. Few of his personal papers survive because of a fire. We know that he finally secured a U.S. patent for his ice machine in 1851 and that two inventors from Europe—possibly having read the slanderous articles—visited his office that same year to see the machine in action. There's also evidence he visited cities in Florida, South Carolina, and New York to search for investors, though where he got the money for travel remains uncertain. He finally found an investor in Boston, but the man died before an arrangement was finalized. No other leads appeared. At his lowest point, he was reduced to walking the streets of New York City, hawking pamphlets about his idea and asking to speak with anyone who'd listen.

In 1855, just four years after receiving his patent, Gorrie sat shivering on his front porch. Bereft of money, friends, and now his health, he would have looked emaciated wrapped in a thick wool blanket. He probably caught malaria—a disease that, like yellow fever, he believed he could cure with ice. When his fever spiked to 104 degrees, he went to bed, where he expired twenty-four hours later, his reputation in tatters and with debts of more than $6,000 (about $200,000 today). He died convinced that Frederic Tudor had destroyed his life.

John Gorrie knew he was ahead of his time. Toward the end of his life he wrote that his ice machine "has been found in advance of the wants of the country."

A decade after his death, the nation's wants finally caught up with his vision. At the start of the Civil War, inventors from France and England simultaneously announced their "discoveries" of the process of manufacturing ice, though their systems were curiously similar to Gorrie's. These were probably the same inventors who had visited Gorrie ten years earlier and who, perhaps, had seen his patents. Free from Tudor's reach, the Europeans improved on the doctor's design by replacing air with ether and ammonia. They sold the designs to manufacturers in the United States in the 1860s, setting in motion America's human-made ice age.

The South in particular embraced ice machines. The war had ended slavery, thus upending the plantation economy that had awarded so much wealth to so few families. The region was now competing with northern industry, and ice manufacturing offered a timely leg up. By 1875, engineers who saw in ice the potential to make them rich opened ice manufacturing plants in Florida, Louisiana, Alabama, and Mississippi. By 1920, the United States boasted more than 4,800 block-ice plants employing 160,000 people and producing 40 million tons of ice per year.

Nearly every hospital in the South built a commercial-sized icehouse on its grounds. Had Gorrie lived, he would have seen one of his biggest dreams come true: hospitals could now treat every patient suffering from surgical wounds, broken bones, or fever with ice.

A medical journal reprinted Gorrie's "On the Prevention of Malarial Diseases" series, this time with his real name, generating interest in his bedpan "cooling machine." The machine's popularity peaked in 1881, when President James Garfield lay dying at the White House from a gunshot wound. His nurses used the apparatus to cool the room and keep the president comfortable.

The cost of ice continued to fall, giving rise to new industries. American merchants shipped ice-packed apples and strawberries to nations around the world. Merchants from those countries shipped citrus to the United States in return. Seeds from Cuban oranges were cultivated into some of the first orange groves in California and Florida. The wider availability of citrus helped eliminate scurvy, a rampant health problem of the day. Apple orchards in New England and across the upper Midwest shipped their products south, where the fruit was less common. Even the poorest of Americans could afford to purchase fresh apples, which they baked into bread and pies. Few other countries had access to such delights, leading international visitors to coin the phrase "as American as apple pie."

Mechanical ice also established the American seafood trade. Fish had been a colonial delicacy on the Gulf Coast for a century—and a staple of Indigenous coastal diets for much longer—but because fish had to be kept cold, it wasn't eaten more than a few miles inland. The affordability of manufactured ice enabled fishermen to pack their cargo hulls with ice, which kept their catch fresh at sea. Back on shore, they packed their still-fresh catch into insulated train cars—another recent invention, thanks to manufactured ice—and shipped it to the Midwest, sparking new appetites for snapper, snook, and mahi-mahi.

The technology changed American sports as well. It was used to create indoor ice rinks, expanding ice skating, one of the most popular pastimes of the late nineteenth century, into urban centers. Ice rinks also helped to popularize hockey and speed skating. Curling likewise found a dedicated fandom in America on human-made ice, with curling clubs cropping up across the country.

And the mysterious crowd poison? It magically disappeared once public spaces began using manufactured ice to cool their buildings. In theaters, ice blocks were placed in front of large fans, which blew air over them to cool auditoriums. When shows were very popular, the fans worked harder and the ice melted more quickly. This likely contributed to critics' use of a word that, by the 1940s, came to describe the most popular shows of the year: "block busters."

Today, John Gorrie's remains rest beneath a tree across the street from his namesake museum. His tombstone is small and easy to overlook. "This is his third resting place," a park ranger named Peggy told me during my visit. "He was first buried by the water, but the sea kept rising, so his family moved him here—or, you know, what was left of him."

A statue of John Gorrie stands in the U.S. Capitol's National Statuary Hall Collection, a gift from the Gorrie Museum, but his story remains mostly untold. The majority of his personal papers were lost to fire, and public records of his work are spread across several hard-to-find archives. A flip through a contemporary HVAC textbook shows that Gorrie's discoveries are usually credited to innovators who based their work on Gorrie's idea, but rarely to Gorrie himself.

There is at least one institution—beyond the museum—that credits Gorrie's achievements. The National Museum of American History in Washington, D.C., houses the doctor's first working ice-making machine prototype and his original patents. At the time of this writing, they remain in storage, out of public view.

CHAPTER 3

The Iceman Cometh

Handsomest Iceman contestant being
powdered by a woman (1940)

"**O**h, I remember him, all right," Lillian told me over the phone. "He was so handsome." Lillian had turned ninety just days before this call. Her rural childhood home, she said, didn't get electricity until the early 1940s. She remembers life before electric light, before vacuum cleaners, before modern refrigerators.

She also remembers the iceman who visited her house every week when she was a girl. She remembers him driving his truck to her house, lifting 50-pound blocks of ice from the back, and carrying them with a pair of tongs into the kitchen, where he'd heave them into the upper part of the icebox.

"He tracked a lot of dirt in," she said. "But mother didn't seem to mind. She thought he was as handsome as I did."

As ice became more popular among working-class Americans, natural and manufactured ice companies competed ruthlessly to woo customers. They promised ice delivery on a weekly basis; all a person had to do was display their ice card in the window, showing how many pounds to deliver.

Icemen were as common in those days as milkmen. Every morning, they filled their wagons—and eventually gas-powered trucks—with blocks of ice, which they delivered door to door in almost every city and county in the country. Icemen were strong and vital, and something about their particular combination of muscle, sweat, and coolness sparked the popular imagination. Consider the 1908 ditty "All She Gets from the Iceman Is Ice," sung by Ada Jones. The singer laments that "there's something in [the iceman's] business that affects his temperature"; he's just too cold, it seems, to give her what she wants.

As Cole Porter's 1948 hit "Too Darn Hot" reminds us, there's a clear correlation between cooler temperatures and an inclination for lovemaking. In the 1932 film *The Dentist*, W. C. Fields plays a well-to-do dentist who reacts with a mixture of shock and disgust to the news that his daughter is marrying the iceman, a burly gent known throughout the neighborhood as a real cad. Eugene O'Neill's 1939 play *The Iceman Cometh* gets its title from a running gag between protagonist Theodore "Hickey" Hickman and his fellow traveling salesmen, who are convinced that one day they'll come home to find their wives "rolling in the hay with the iceman." Newspaper comics leaned in to the sexed-up iceman stereotype. One recurring gag was to have an older husband—presumably too old to be much fun in bed—greet the wide-eyed iceman in the kitchen, saying something like, "The wife isn't here—just give me what she always gets."

Even the *Saturday Evening Post*, a newspaper known for its wholesome content, published a short story in 1954 called "Judy and the Iceman." It ran alongside an illustration of a teen with red-painted lips surprising an iceman with a stolen kiss. As a writer for the *Washington Evening Star* once put it, the iceman is "the theme of song and story," especially romances.

But on what was such a reputation built?

The ice historian Jonathan Rees poses one theory. "[Icemen] were symbols of male virility," he writes, "particularly since they often visited housewives while their husbands were out working." Unlike the milkman, who left his goods at the doorstep and went on his way, icemen came into homes to put ice directly into iceboxes. The work was strenuous, and it wasn't uncommon for a housewife to offer her iceman a refreshing glass of water or a bite to eat, perhaps a chair on which to rest a moment. By entering the private, domestic spheres of housewives—something few other deliverymen did—icemen were seen by some as crossing a forbidden threshold.

The fantasy was fed by the iceman's typical brawniness. These were men who could—because they had to—balance a 50-pound block of ice on their shoulder while dangling another from a pair of metal tongs. In major cities such as New York and Washington, D.C., icemen would carry both blocks simultaneously up six flights of stairs or more, an impressive display of "virility" if ever there was one. Perhaps history's most famous iceman was the football legend Harold "Red" Grange of the Chicago Bears, who delivered ice in summer to stay in shape.

Icemen appeared in more than romances, however; they often made cameos in comedies, too. In one of the most famous episodes of *The Three Stooges*, the inept trio try their hand at ice delivery, only to have the blocks of ice melt to the size of cubes before they get them up the stairs. In *The Girl Can't Help It*, an iceman spots curvaceous Jayne Mansfield as he's lifting a block of ice from his truck. The sight of her raises his temperature, melting the block into steam. In *Horse Feathers*, Chico Marx plays a highly sexed iceman who chases Thelma Todd, a routine powered with something more than slapstick silliness. The cultural critic Christopher Miller has argued that "unlike the milkman, as pasteurized and homogenized as his ware, the stereotypical iceman, at least in eastern cities, was a recent Italian immigrant." Thus, his ethnicity turned him into someone not only to joke about—but to fear."

In the heyday of ice delivery, icemen would visit homes about once per week. By 1895, approximately three thousand horses were pulling 1,500 ice wagons in New York City and the still-separate city of Brooklyn, which together formed the largest market of consumable ice in the United States. Ice wagons working for natural-ice companies filled up at piers, where icehouses stored blocks of ice shipped from Maine or the northern Hudson River tributaries. Wagons for mechanical-ice companies filled up at plants that operated at the edges of the cities.

Some wagons could hold a few hundred pounds of ice, others up to several tons. Each wagon's braced sides were connected by a canvas top brandishing the name and logo of the ice company that owned it. Most were painted white or yellow to refract the sun and keep the ice as cool as possible. Over time, the bright colors faded from dust and sun, leading one newspaper to describe ice wagons as "lurid" and something out of a "defunct circus."

Ice delivery was dirty business. Wagons carrying natural ice were streaked at the end of the day with dirt and sediment that had been frozen in the blocks. This was especially true for wagons that delivered to bars and taverns, which were renowned for buying the dirtiest ice they could get, because dirty ice was cheapest. Icemen got filthy, too. They worked hardest in summer, when demand was high and the heat was tortuous, dripping sweat and tracking dust and grime into kitchens. The horses suffered as well. It wasn't uncommon during the hottest weeks of the year to see the animals straining then falling dead in the street. Their corpses would sometimes lie for a day or more before local authorities collected them.

The ice business was also dirty in the metaphorical sense. The least scrupulous of icemen sold a portion of their company ice on the side, pocketing the proceeds and blaming the missing ice on inevitable melt. For years, ice companies took their icemen's word that they were conducting business on the level, because there wasn't an accurate method in place to track inventory. But by the 1910s, the companies had lost too much money to rely on the honor system. Coupon books were distributed to customers for a fee. They were told to exchange coupons with their icemen instead of money. The icemen, meanwhile, were forced to create detailed records of their deliveries, an accounting system that ushered in the sort that commercial truck drivers are required to use today.

In those pivotal years between 1880 and 1920, the ice industry was changing fast. Ice had become one of the biggest industries in the country. It was also one of the biggest indicators that something had happened to the lakes and rivers from which it was harvested.

———————————

Philadelphia, 1883. Early morning. The sun rose behind the skyline, throwing into silhouette the towering smokestacks that blew clouds of ash and smoke into the atmosphere. Toxic waste from factories poured into the Schuylkill River, the primary source of Philadelphia's drinking water. On the Schuylkill's west bank, just north of Market Street, hundreds of cattle, pigs, and sheep ran frenzied through a slaughterhouse gate, where they met a gruesome end. The blood and detritus from their slaughter became a black tributary that poured into the river.

That same year, a local newspaper described the Schuylkill as having turned "a dark greenish color," with "thousands of dead fish... lying about the banks." It wasn't uncommon to witness bits of wool and cotton fabrics created in nearby mills floating past, seeping dye into the water. A chemical engineer hired by the city named Julius W. Adams concluded that the river was "not a proper water" for human consumption, and Philadelphia banned ice harvesting in its immediate proximity. Ice companies got around these restrictions by harvesting just a short ways to the north, where the water wasn't much cleaner, but the ban couldn't reach them.

Cities that reported problems with their water also found problems with their ice. A chemist working in New York City in 1907 examined ice harvested from the Hudson and declared it a "health hazard." This ice, his report said, was "unfit for human consumption." Of the twenty-five samples taken from the river, most contained what he called "intestinal germs." Germ theory was still new, and no one really understood what germs were, so the scientist categorized the samples according to smell, which ranged from "vegetable" to "disagreeable" to "rank."

In upstate New York, the St. Lawrence State Hospital authorized the cutting of ice from the St. Lawrence River just a few hundred yards from the building's main sewer line. The hospital suffered a typhoid fever outbreak that resulted in several deaths among patients and staff. Given the hospital's relatively secluded location, investigators were able to quickly trace the outbreak to ice taken from the river.

In Detroit, ice consumers fell ill at record rates from waste that gas companies leaked into the Detroit River and Lake St. Clair. A writer for the *Detroit Tribune* argued, "It is probable that many of the summer ailments of the digestive tract are due to the use of villainously impure ice, which is loaded with dormant bacteria." In Chicago, a similar problem forced ice companies to move their harvesting away from Lake Michigan to the somewhat purer waters of Wisconsin's Fox River.

Across the country, the public's confidence in the safety of natural ice plummeted. Ice cards printed that year by the Knickerbocker Ice Company, the largest natural-ice company in Philadelphia, reveal just how strongly customers there and in New York City rejected their local product. Stamped diagonally across every card were the words

"We furnish PURE EASTERN ICE ONLY," a promise that the public would take to mean that the ice had been cut and hauled from distant, unspoiled Maine. Pollution had become so bad, and customers so wary of contaminated ice, that even the country's largest natural-ice company had to seek purer sources of ice or risk going out of business.

In 1906, the federal government passed the Pure Food and Drug Act, which was supposed to protect consumers from contaminated food, but it did little to regulate the ice industry. The only penalty an ice company suffered for selling dirty ice was a $150 fine—not much of a deterrent for an industry raking in nearly $1 billion per year.

Manufactured-ice companies, meanwhile, knew a great marketing opportunity when they saw one. Some companies publicized photos of dolls or toy cars frozen in the center of their blocks of ice to demonstrate their purity. When Mark Twain visited an ice plant in New Orleans, he marveled at how objects frozen in blocks "could be seen as through plate glass." Other companies hired experts to expound upon the dangers of tasting natural ice. An ad from 1914 featured a testimonial from a physician: "The character of the lake ice which has been found in my own ice chest has been bad."

Natural-ice companies and their allies responded first by evoking Gorrie's critics from a half-century earlier. "The difference between manufactured ice and natural ice is that one is a live product of the Creator," wrote a *Chicago Chronicle* reporter in 1903, "while the other is the dead product of mere man and his little machine." When that didn't work, they addressed the pollution issue with amusing directness. Read one ad by a natural-ice company in Ballston Spa, New York: "No sewage."

Water pollution wasn't the only problem natural-ice companies were facing. The Little Ice Age, a naturally occurring period of planetary cooling that had for centuries produced extremely cold winters, had come to an end. Just eighty years after the industry took off, the thick blocks of ice that Frederic Tudor's men had once cut from Massachusetts lakes were now a thing of the past. Lakes and rivers simply didn't freeze as deep as they used to; some years, they hardly froze at all. During warm winters—or "open winters," as they were called by industry insiders—the press would warn of a pending ice shortage, and the price of ice would skyrocket.

Taken together, these changes reshaped the public's view of mechanical ice. By the end of the nineteenth century, such ice accounted for nearly 15 percent of the country's total consumption, and the cries of blasphemy that had once ruined John Gorrie's reputation were no longer heard. The natural-ice industry was in a downward spiral. One can imagine CEOs of natural-ice companies pacing behind closed doors as their accountants warned of sagging profits. In 1913, the Natural Ice Association of America published a pamphlet called *The Handwriting on the Wall.* Subtitled *A Call to Arms!,* it warned natural-ice dealers that manufactured ice would soon destroy their empire. "The time has arrived," the publication read, "for definite action on your part. Our work will be to educate the public. . . . You owe it to yourself to join forces with us all to preserve our markets in the years to come."

This plan to "educate the public" included a peculiar marketing campaign to recast natural ice as a luxury item, and it worked, for a while at least, among the metropolitan wealthy. A 1906 article in the *Washington Evening Star* reported that "manufactured ice is not saleable in some parts of Gotham at any price." Even hotels with their own ice plants were "compelled to buy Kennebec or Hudson River ice to supply their patrons who are particular."

The boost in sales was temporary, however, and the natural-ice industry continued to decline. Executives understood that the only way their natural-ice companies would survive was through consolidation. Across the country, regional ice companies merged, forming monopolies and trusts headquartered in the biggest cities. Many were run by robber barons, who squeezed as much profit as possible out of a dying industry. The most notorious was Charles W. Morse, president of the New York City "Ice Trust," whose selfish pursuits led to the deaths of more than a thousand New Yorkers.

I n 1890, New York City was consuming more than 285 million tons of ice per year, more than anywhere else in the United States. Natural ice still led the industry, but mechanical-ice plants were cropping up along the Hudson. The city's largest supplier of natural ice was the Knickerbocker Ice Company, which had secured its place in the market by harvesting ice from Maine, where the lakes were secluded and streams still ran mostly clear. By 1891, the company was running sixty barges a week, each carrying 900 tons of ice, from Maine and the northern parts of the Hudson to New York's five boroughs.

Nobody had more people cutting ice than Knickerbocker. Nobody, that is, until the arrival of Charles W. Morse, a wealthy businessman from Maine who'd made his fortune in the shipbuilding industry. Seeing potential in the ice market, Morse bought up smaller, regional natural-ice companies to form the Consolidated Ice Company. The company devoured so many little competitors that it was eventually able to outsell Knickerbocker in a single year. And then, it gobbled up Knickerbocker, too.

Now owning more than half the natural-ice market in New York City, Morse hiked up prices by 100 percent and blamed the rise on an ice shortage. The city's wealthiest residents could absorb the hike, but it devastated the city's poorest. They couldn't afford ice, and without ice they couldn't preserve food or milk. Worse, they couldn't get cool.

An article in the *New York World* warned that "unless something is done in the matter there will be widespread suffering among the poor people and a higher death rate than ever among the sick babies of the tenements during the hot weather."

The warning was, unfortunately, spot-on. In the summer of 1896, a deadly heat wave struck New York City. Temperatures rose to the nineties and stayed there for over a week. The city's poor—most of them immigrants—lived in tight, small brick tenements on the Lower East Side, which during the day became ovens with temperatures inside reaching nearly 120 degrees. The *New York Herald* described people sleeping virtually everywhere to stay as cool as possible: "The sidewalks were lined with men, women, and children. Cellar doors were their mattresses and beds were made in trucks and wagons." After the heat wave's first night, the sun rose on a funereal scene: "From block to block long rows of baby carriages filled the gutters, and from street to street there went a wail of misery and discomfort."

Nearly a week of deadly heat passed before the mayor, William L. Strong, formed a task force to determine how best to bring relief to the city. This was a time when small-government ideology dominated political discourse. Elections were won by men who argued the loudest for the least amount of government intervention. So, in this case, the government elected not to intervene at all, despite evidence everywhere that the city was in crisis. Not even the state governor, Roswell P. Flower, sent help. "It is not the providence of the government to support the people," he scoffed at a press meeting several days later.

By the time the heat dissipated, 1,300 New Yorkers had died, more than ten times the number that perished in the 1863 New York City draft riots and four times the losses in the 1871 Great Chicago Fire. Many more would have died, too, if not for the swift thinking of a New

York City police commissioner, who took action while the mayor and governor did nothing. The commissioner ordered hundreds of truckloads of ice—natural and mechanical, whatever could be spared—to be delivered to the city's poorest neighborhoods. When the trucks arrived, small children leapt onto the tailgates and broke off chunks of ice to toss to their families and neighbors. Dozens of icemen and the commissioner himself unloaded the free ice and carried it into tenements, where citizens cried with relief and offered grateful handshakes.

In the months that followed, the people saved by the commissioner's ice delivery became dedicated supporters of his political ambitions, giving meteoric rise to his career as a politician. The commissioner was Theodore Roosevelt, and in just five years, he would be president of the United States.

In a letter to his sister, Roosevelt described the heat wave of '96 as "the worst and most fatal we have ever known." He continued: "We had to distribute hundreds of tons of ice from the station-houses to the people of the poorer precincts" because "the death rate trembled until it approached the ratio of a cholera epidemic."

When the heat wave finally ended, the newspaper publisher William Randolph Hearst encouraged an investigation into the Consolidated Ice Company, which, in the eyes of the public, was at least as responsible as the mayor and governor for the deaths of so many New Yorkers. That investigation inspired lawmakers to pass anti-monopoly legislation, but it would be years before the laws were implemented effectively. In the meantime, citizens had to pay up—or no ice.

Morse continued to build his frozen empire. By 1899 he had bought most of the mechanical-ice plants in Manhattan and Brooklyn, which were his only competitors left in the region. He now owned virtually all the ice coming into or being made in the greater New York City area, and he was expanding into Philadelphia and Washington, D.C.

He renamed his growing empire the American Ice Company, which newspapers called "the Ice Trust."

To ensure his monopoly, Morse made it impossible for the ex-CEOs of the ice companies he bought out to ever compete with him again. He obtained a permanent injunction from the New York Supreme Court restraining them from working in the ice trade, and anyone who challenged the injunction quickly learned their lesson. According to a story in the *New York Times*, a bought-out ice dealer named W. A. Wynne tried to start a new ice company in the city from scratch, only to awake one morning to a steamer hired by the American Ice Company smashing all the ice in front of his business.

Over the next year, Morse continued to raise prices and blame the hikes on ice shortages. His excuses made little sense to anyone outside New York City. In Buffalo, Boston, and Albany, ice prices remained steady, and supply was easily keeping up with demand. Only in New York and Philadelphia, where the Ice Trust was strong, were prices higher than what most people could afford. The *New York Times* launched its own investigation into Morse and within a year ran an exposé that proved that the ice shortages were a lie. It further linked Morse's shady business practices to the city's new mayor, Robert Anderson Van Wyck, and the powerful and increasingly corrupt political machine, Tammany Hall.

"ONE HUNDRED PER CENT RISE IN ICE" read the paper's all-caps, six-column headline. The article revealed that while Morse's company claimed it had harvested only 1.4 million tons of ice the previous winter, it had actually harvested more than 4 million tons—more than enough to supply a desperate city. Moreover, the mechanical-ice plants produced an additional 700,000 tons. The *Times* further revealed that despite the price hike (or rather because of it), the price of American Ice Company stock was going through the roof, and the people profiting most were some of the highest-ranking political officials in the city. Van Wyck, who made $15,000 a year as mayor, was in possession of American Ice Company stock worth $680,000. The Tammany boss Richard Croker held nearly a quarter of a million dollars in stock in his wife's name, and his underboss, John Carroll, held a half million. The Hudson dock commissioner Charles F. Murphy, who would one day take over from Croker, also owned a large chunk of stock. Not coincidentally, he and his men had been in charge of enforcing laws that kept other ice companies from trading in the city.

The *New York Times* was silent for three weeks after these revelations until it ran a series of damning articles on the trust. Other newspapers followed suit. The *New York Evening Post* ran articles explaining that ice plants could be constructed to produce ice that could be sold for a mere ten cents per 100 pounds, ensuring that there'd never be a faux ice shortage again. The Hearst-owned *New York Journal* dropped another bombshell in the form of a list of public officials who owned American Ice Company stock and what each of their shares was worth. Perhaps most damning was the newspaper's revelation that Van Wyck hadn't paid a cent for his shares.

The articles caused a sensation in the city, destroying the careers of several Tammany men and politicians at every level of government. Croker let John Carroll take the fall and publicly let him go. Mayor Van Wyck lost his bid for reelection to a reform candidate in 1901 and watched his political career disappear overnight. The mayor's brother, Augustus, a former judge, also made the list and subsequently lost his run for governor of New York to a candidate also associated with ice, but for very different reasons. He lost to Theodore Roosevelt.

The fall of so many political figures did little to stanch public outcry. The *Times* reported that the trust "is talked of everywhere, from the slums to the clubs, in Wall Street, and on street cars." In general, the

Times concluded, the trust is "now loathed by the community." To be clear, this wasn't just antitrust sentiment. In fact, the *Times* went out of its way to explain it was averse only to "bad" trusts, calling the Ice Trust a "hoggish monopoly." The Ice Trust's problem, as the press and public saw it, was twofold: First, it was clearly corrupt. And second, at the turn of the twentieth century, ice was no longer a mere luxury—it was a utility essential for staying cool and preserving food. "To corner ice," wrote the *Times*, "is very much like cornering air and water."

Over the next few years, the American Ice Company felt the intense watch of journalists and lawmakers, and so kept the price of ice steady. Then, in 1910, its largest facility, in Iceboro, Maine, burned to the ground, turning the company's ice schooners to ash. This had been one of the last remaining ice fields safe for human consumption in the country, so natural-ice delivery effectively came to a halt in the city. One year later, the company split up to avoid further legal action, becoming the Knickerbocker Ice Co. and the Ice Manufacturing Co.

But what of New York's newest ice king, Charles Morse?

During the scandal, most stockholders lost their money, their careers, or both. But not New York's slipperiest snow mogul. It was soon revealed that Morse had silently withdrawn from the American Ice Company in 1901 with over $12 million in his pocket. Looking for new ways to make money, he turned to the shipping business and by 1907 had managed to quietly build another monopoly in shipping from Bangor, Maine, to Galveston, Texas. To industry insiders, Morse had become the "Admiral of the Atlantic."

Morse wasn't finished wreaking havoc, however. In October 1907, he joined forces with F. Augustus Heinze, president of the Mercantile National Bank and founder of United Copper, to corner the market on United Copper stock. Their scheme failed, triggering a panic on Wall Street that led to runs on the banks associated with Heinze and another friend of Morse's, the president of the Knickerbocker Trust Company. The banks were forced to close, and the New York Stock Exchange fell by more than 24 percent in the first nine months of the year. The banker J. P. Morgan stepped in and stabilized the economy by pledging his own money, but the panic continued to cause disruptions throughout the financial sector. The eventual devastation was the worst economic crisis the country had seen, later eclipsed only by the Great Depression.

The Panic of 1907 settled down the next year, but once again public outcry led to an investigation of Morse. This time, he faced an indictment and conviction for tax fraud, of all things. (His shipping company had failed to pay its fair share of taxes for years.) An article in the *New York Times* described Morse at his trial as a "fat, squatty little man" with "masterful inquiring eyes." The fallen ice king was unrepentant, however. "There is no one in Wall Street," he said at the trial, "who is not daily doing as I have done."

Morse was sentenced to fifteen years in an Atlanta penitentiary, where he seeded yet another criminal legacy. While there, he umpired a prison baseball game that included an inmate who worked for a New York City mob family. The man's cellmate was a short, dapper man named Carlo, who'd been arrested for smuggling Italian immigrants into the country. Morse took Carlo under his wing, becoming the younger man's mentor. Whatever wisdom Morse may have imparted to his young protégé is lost to history, but a few years after his release, Carlo, now known as Charles Ponzi, began illegally selling international reply coupons through the U.S. Postal Service. The scam became one of the largest and most infamous pyramid schemes of all time.

Over the next couple years, Morse's friends and relatives attempted to secure a pardon, but President William Howard Taft, recognizing the man's pattern of criminal behavior, refused. Then, in 1912, a lawyer named Harry Daugherty secured Morse's release. (Daugherty would later serve as the attorney general for President Warren G. Harding, considered by many historians to be the most corrupt American president in history.) Morse, it seems, had fallen mysteriously ill, and Daugherty convinced doctors to examine him. Given Morse's symptoms, the doctors concluded that he was suffering from Bright's disease and would not last a year. Taft reluctantly signed a pardon, and Morse was let go.

A year later, Morse was still alive (and Daugherty's fee remained unpaid). Taft's attorney general's office received a tip that, just before his examination, Morse had drunk a combination of soapsuds and chemicals to produce the temporary symptoms of Bright's. Taft proclaimed he'd been duped, adding that the case "shakes one's faith in expert examination." But a presidential pardon, once accepted, cannot be revoked.

In 1922, Morse made the news again. His newly formed United States Shipping Company had prospered during World War I but was now under investigation for war fraud. Morse was indicted a second time, but before the case could be brought, he was indicted once more on the charge of using the mail to defraud potential investors. Before either case could be tried, Morse was judged too ill by a physician hired by his lawyer to stand trial and was placed under guardianship. Yet again, once his legal danger passed, his health endured. He died eleven years later in his hometown of Bath, Maine, at the age of seventy-six.

Like the original Ice King, Frederic Tudor, Charles W. Morse died a very wealthy man. But whereas Tudor earned his wealth by bringing ice to Americans, Morse built his fortune by denying it to them. In less than one hundred years, ice had become so vital to American life that it had accrued the power to take lives, build fortunes, and destroy careers.

———————————

In the decade before Morse's death, the natural-ice industry made one last gasp. In 1917, the United States entered World War I. By now, mechanical-ice plants had upgraded their ice machines to use ammonia to freeze water. These new ice machines were a major improvement over earlier technologies, producing thicker and stronger ice that could be shipped greater distances and stored for longer periods of time. But as the war raged on, ammonia, a key ingredient in the production of munitions, was in short supply. The shortage caused ice plants to halt production and furlough their workers. This, in turn, cut off the supply of mechanical ice that Americans had come to rely on. The National Association of Ice Industries teamed up with the federal government and launched a national campaign to bring men back to the ice fields.

Across the country, pasted to the sides of buildings and hanging in business windows, were brightly colored posters asking men to "help in the harvest" because "ice is needed to save food for the starving people of the world." Out-of-work ice-plant employees joined farmers and other laborers looking for seasonal work on the ice fields in Maine and upstate New York and throughout the upper Midwest. For almost three years, natural-ice harvesting once again became the primary source of ice for Americans. And ice delivery, once the domain of men so strapping that the profession was often viewed as the height of masculinity, became one of the first industries to employ women, as men left to fight overseas.

Ice deliverywomen, often working in pairs, loaded 50- and

100-pound ice blocks onto delivery trucks, drove to the homes of customers, and hauled the ice into customers' kitchens. As for the few icemen still making deliveries, they found themselves once again the "theme of song and story" as rumors spread that they were enjoying too much the company of home-alone housewives. Returning-soldier husbands were determined to get rid of iceboxes—and the icemen who filled them.

The blues singer Casey Bill Weldon's 1936 song "We Gonna Move to the Outskirts of Town," perhaps best known today from Ray Charles's 1961 cover, captures the anxiety of jealous soldiers returning home:

I don't need no iceman, 'cause I'm gonna get me a Frigidaire.

Within a few years of the war ending, the electric refrigerator went from being a novelty of the rich to one of the country's most common household appliances.

CHAPTER 4

Highways, Holidays, and
the Cheap-Ice Revolution

People gathered around pies on a Coca-Cola cooler
during the wheat harvest in Nebraska (1959)

On a hot summer evening in 2018, while visiting family in Kansas, I received a call from an old high school friend who invited me over; she asked if I'd bring a bag of ice. Kansas summers are best spent sitting outside with a cooler full of ice and cold beer. My thoughts thick from the heat, her request completely slipped my mind until I was well on the road. She didn't live far from a 7-Eleven, so I pulled in.

A bell chimed when I opened the door. I waved to the clerk and pointed to the white metal box out front that had the word "ice" painted on it in tall red letters. The clerk rang me up while I glanced around at what was for sale: a plastic cup holder that fits in the driver's side window, a Styrofoam cooler for last-minute road trips. It occurred to me that I rarely—if ever—see bags of ice for sale in my local New York City bodega. Instead, you find it in places like this, by the bag or the cup, in gas stations, convenience stores, and large suburban grocery stores, where folks arrive more often by car than on foot.

That makes practical sense, of course. You can toss a cooler into the trunk of your car, but lugging your own bag of ice several blocks and up four stories? That's a level of hassle even New Yorkers won't put up with.

It also makes historical sense. As the twentieth century moved on, finding a nation in motion, America's obsession with ice was in many ways related to its obsession with cars, with the open road, and with the very idea of convenience.

T he Roaring Twenties ushered in an age of cultural transformation. The Great War was over, and in America, World War II was still a specter of conflict in Western Europe. The women who'd delivered ice while men were fighting overseas wanted to keep working and sought careers outside the home. Their fashion sense changed along with their ambitions. They cut their hair short and raised their hemlines. Men, meanwhile, shaved their beards and traded formal day wear for sports clothes. The Civil War, now in the relatively distant past, had rendered slavery illegal in America, but racism was still prevalent and Jim Crow laws enforced segregation in the South. It was during this time that the first of more than six million Black Americans left the South for northern cities like Chicago, Detroit, New York, and Philadelphia. Many of them went to work at the Ford automobile plant in Detroit, one of the few factories in the country that paid white and Black workers the same. America was a country in flux, and the ice industry was evolving with it.

In the North, especially in cities, people still relied on the iceman for weekly deliveries. But in the South and Midwest, where towns stretched for miles and middle-class families were quickly becoming car owners, it wasn't uncommon for folks to stop at an ice dock on their way home after work to purchase a block or two. Ice docks were small wooden buildings where retail ice companies stored their supply, either shipped in by train or created in a nearby ice plant.

The Southland Ice Company in Dallas, Texas, was one of the first to take notice of this trend. The company formed during the winter of 1927 under the management of Joe C. Thompson Jr., described by his colleagues as someone who "worshipped an 'idea.'" He believed that during times of extraordinary change, businesses had to change, too. The company opened dozens of ice docks throughout Texas, one of which was run by John "Uncle Johnny" Green, who came to the ice industry from the theater business, a line of work that had taught him that spectacle sells. He designed his dock's front display to attract the eyes of people driving by: red flags hung from the roof, fresh green watermelons glistened out front, and colorful painted signs hung in the window to brighten up the otherwise gray building.

Green noticed that many customers stopped by the dock after doing their grocery shopping at the local seller or the A&P, then America's largest grocery chain, with around fourteen thousand locations. He overheard them grumble about forgetting ingredients and feeling frustrated at the thought of having to drive back to the store to pick them up. It occurred to him that if he stocked up on kitchen staples like milk, eggs, and bread, he could make extra money off his customers' forgetfulness. His strategy worked even better than expected. He attracted so many customers, he had to extend his hours to Sunday mornings to keep up with demand.

Customers loved it. Ice and groceries in the same place, and they could buy what they needed on their way to and from church. Less happy were the grocery stores. They had long kept closed on Sundays out of tradition and religious observance, and they resented the business they were losing to Green. Store owners called Green's hours "sacrilege," and his business practices "dirty pool." But the complaints did nothing to dissuade Green from his new business model.

Green's success attracted the notice of Thompson, who saw the financial potential of Green's ideas and replicated them at Southland's other ice docks throughout Texas. When sales picked up at those, Thompson asked his dock managers to provide customers with paper tote bags so that they could better fit their groceries in the trunks of their cars. The bags were so popular that Thompson rebranded the docks as Tote'm Stores. Playing with the name, he purchased totem poles from an Indigenous carver in Alaska and installed them in front of every store.

At a convention for ice industry executives in 1929, Thompson delivered an address that predicted car ownership would only increase Americans' use of ice, and so he urged his colleagues to do whatever they could to attract car owners. For his part, Thompson installed gas machines in lots next to his ice docks, creating the first outline of what would become the familiar gas station–convenience store combo that would go on to shape the landscape Americans use today.

Thompson's prediction was off in some particulars, however. Americans would increasingly rely on ice in their homes, but car owners wouldn't always have to drive for it. Just ten years earlier, a new automobile conglomerate called General Motors had bought Frigidaire and had begun mass-producing refrigerators on its assembly lines. As the cost of fridges came down, more people than ever before—especially those who lived in cities where electricity was plentiful—purchased them for their homes, sending their old iceboxes to the curb. These newfangled fridges could seemingly do it all: they could keep perishables cold and even make their own ice. What's more, they didn't leak meltwater the way even the best-constructed iceboxes did.

A sales document from the era reveals that ice companies did everything they could by way of marketing to stifle interest in this new technology, including becoming among the first in the country to hire women as salespeople, pioneering the woman-to-woman approach of companies like Avon and Mary Kay. These saleswomen would visit homemakers during the day with icebox brochures and memorized pitches about the benefits of fresh ice. Sometimes they'd wink and mention the iceman.

In the 1930s, General Electric began manufacturing cheaper

refrigerators that the average American could afford, disrupting the market that Tudor had created. By the end of the decade, most of the old-fashioned iceboxes still in use in America were owned by urban families living below the poverty line or by those in rural areas. Then, thanks to the New Deal's Rural Electrification Act, which finally brought electricity to the most remote parts of the country, rural homes began to embrace refrigerators, too.

The ice industry might have collapsed with the rise of refrigeration if not for America entering World War II. Soldier camps required an enormous amount of ice to maintain the health and happiness of their enlisted men, and the military called on the mechanical-ice companies to keep them in supply. One of the biggest suppliers was the Southland Ice Company. Its mechanical-ice plant in Sherman, Texas, was tasked by the federal government to serve Camp Hood, one of the U.S. Army's largest training camps in central Texas. The plant's first contract with the army called for 32,000 tons of ice to be delivered each week, more than twice as much as the plant had ever produced previously. The order strained the bodies of the men working there. According to E. H. Roberts, who operated the Sherman plant until 1942, the workers "did not have a lot of help because most of the men had gone to war. Trucks were running all over Texas locating enough ice to fill the order."

If a delivery fell behind schedule, the plant received an angry call from an army officer. "I virtually never slept at home in those days," said Roberts. "I would sleep in the ice plant on top of sacks of salt." The Sherman plant teamed up with another Southland plant in Temple, Texas, but the work stayed grueling. In a 1977 interview, a Temple plant worker named Wiley Wesson said that during the busiest months of the war, he carried "about a hundred 300-pound blocks of ice" every twenty-four hours.

At their operational peak, the Sherman and Temple plants were running more than three hundred delivery trucks per day. "It was a 60-mile round trip from Temple, so that meant two trips before lunch and two trips after lunch," Wesson said. "I guess we loaded and unloaded 30,000 pounds of ice each day. When I got drafted and went into basic training it was like a vacation for me." Wesson was one of only a few icemen that year who went to war. By 1942, ice plants had become so vital to soldiers that a man could be excused from active duty if he worked in one.

By the time the war ended, the Southland Ice Company had made millions of dollars, but it and ice manufacturing companies everywhere were no longer required to supply ice to soldiers and so found themselves transitioning back into peacetime production. Electric refrigerators had caught on in a big way, and ice delivery was becoming an artifact of the past, with smaller ice companies across the country closing their doors. The few that remained open primarily served commercial industries like meatpacking, fishing, and produce transport. Others began selling ice in wax paper bags to compete with household ice makers, and the packaged ice industry was born. The International Packaged Ice Association (IPIA) was formed to support those companies by finding new ways to market ice. That decade, the organization revitalized its 1930s-era "America's Handsomest Iceman Contest." If anyone could sell ice, it was a handsome iceman. It was run like a beauty pageant, with icemen posing coyly and demonstrating their strength by lifting blocks of ice with their bare hands.

In the 1960s, the IPIA battle against homemade ice began in earnest. The organization launched a national publicity campaign that warned the public of the dangers of eating bacteria-laden homemade ice. They placed ads for packaged ice in *Good Housekeeping* and premiered a jaunty new mascot named Chilly Billy. Chilly had an ice cube for a head, a stickman-like body that resembled plumbing pipe, and circles for hands and feet that vaguely resembled ice cubes. In Billy's right hand was a rectangle labeled "ice." The image did little to endear people to the product.

The 1970s brought another major marketing push. The first glass-door refrigerators appeared in grocery stores, liquor stores, and gas stations, and plastic bags replaced those made of wax. Ice companies distributed coupons to potential customers that explained why packaged ice was cleaner than the ice they made at home, and to a degree, the campaign worked—packaged-ice sales increased steadily, until it became the go-to way to get ice for parties and camping trips. By the 1990s, ice plants were fully automated, producing thousands of bags of ice per day. The savvier fast-food restaurants saw a sales opportunity and began selling bags of ice. McDonald's, Arby's, Dunkin', and Burger King still sell it today, and Sonic's packaged "pellet" ice has become so popular that small machines replicating its size and texture can be purchased for use at home.

It was during this half century of transition that Thompson, wanting to ensure Southland's survival, surveyed what was left of his Tote'm Stores and realized that because the stores had all served as ice docks, they shared the following characteristics: they featured open fronts that made curbside service for car owners possible; they were open early in the morning and late at night (a rarity in those days); and they sold ice (by now, mostly in bags) as well as groceries. No other store model in America looked like this, so why not emphasize its uniqueness? He rebranded the stores to reflect their hours of operation, from seven in the morning to eleven o'clock at night, even on Sundays, and changed their name to 7-Eleven.

Over the next twenty years, other companies modeled their version of the "convenience store" on Thompson's 7-Eleven. They mushroomed up throughout the South and, by the 1960s, through the North as well. They became especially popular in America's first suburbs, where cars were necessary to get anywhere and convenience mattered more than almost anything else. Other innovations followed. The 7-Eleven chain claims to have invented self-serve soda fountains and the "to-go" cup of coffee designed specifically for drivers. The store's most famous creation, however, harks back to its origins as an ice company.

In the 1950s, a farmer from Barnes, Kansas, named Omar Knedlik opened a Dairy Queen. On especially hot days, he stored bottles of root beer in his freezer, half freezing the sodas and selling them as "the coldest drink in town." His customers loved the icy texture so much that he sought to make a machine that could produce the frozen drink in half the time. He succeeded using a homemade air conditioner built from parts of an old automobile and started calling the drinks Icees. In 1966, 7-Eleven purchased his machine and the rights to the Icee name, and with that, the Slurpee was born.

———————

Ice and cars remained deeply entwined throughout the twentieth century. The 1950s brought the country's first interstate system and, along with it, a stream of written accounts of life on the road. Jack Kerouac's *On the Road* hit shelves in 1957, John Steinbeck's *Travels with Charley* came out in 1962, and Peter S. Beagle (who wrote *The Last Unicorn*) published his scooter travelogue, *I See by My Outfit*, in 1964. Movies like James Dean's *Rebel without a Cause* hit the big screen in 1955, and throughout the decade, car-themed songs like Chuck Berry's "Maybellene" and Jackie Brenston's "Rocket 88" filled the radio airwaves.

America's proliferating car culture gave rise to another American phenomenon: the road trip. Ads for cars helped sell the road trip to families by suggesting it was the "true American" way to bond with each other. Philip Delaney, an early observer of car culture, rhapsodized about the road trip in an op-ed for his local newspaper: "When [the automobilist] is tired of the old, there are new paths to be made. He has no beaten track to follow, no schedule to meet, no other train to consider; but he can go with the speed of an express straight into the heart of an unknown land." Such romanticizing of cars and life on the road elevated the road trip to something like an act of manifest destiny: by car, Americans would conquer the land, driving on interstates that cut into what had once been forests and prairie land to realize their dreams of seeing the country from sea to shining sea.

For many Americans, the road trip became a kind of rite of passage. The problem that many road-trippers encountered, though, was a lack of ice. While traveling by car was still novel in the 1950s, the use of ice wasn't. By 1951, approximately 80 percent of Americans owned electric refrigerators, many with built-in ice makers. Ice had become a feature of everyday life, and the thought of spending days on end without it kept some cold-loving Americans from taking to the road. Their hesitancy sparked an idea in a man named Richard C. Laramy from Joliet, Illinois, who in 1953 patented the first "portable ice chest," a contraption known today as the "cooler." By the early 1960s, cars and coolers went hand in hand. If you had one, you almost certainly had the other.

Yet it would take more than the invention of the cooler to encourage ice-loving Americans to road-trip for pleasure, because it was nearly impossible to keep a cooler full of ice on the road. Just as Frederic Tudor had learned when he tried to sell ice for the first time in the tropics, mid-twentieth-century Americans were realizing that infrastructure was needed to ensure a steady supply of ice on the road. It would be another decade before the 7-Eleven type of convenience store became commonplace, and grocery stores at this point in history rarely sold ice. Restaurants served ice to customers only in drinks, and roadside inns offered little more than four walls, a roof, and a stiff bed that creaked loudly with every turn in the night.

Among the many Americans frustrated by this lack of ice was the Memphis ice-cream parlor owner Charles Kemmons Wilson Jr., who in 1951 packed up his car and headed out on a road trip with his wife, Dorothy, and their five children to see the sights in Washington, D.C. As the trip wore on, he felt increasingly frustrated that he couldn't replenish the ice in the wooden box they'd packed to hold their lunches. Moreover, he hated paying a surcharge for every child at every roadside lodge in which they stayed. By the time they returned to Memphis, Wilson had come up with the idea to open hundreds of uniform lodgings across the country that would provide comfortable beds and working telephones in every room. And crucially, every floor would have an icebox that provided free ice. He realized his dream the next year, in 1952, when he opened the first Holiday Inn in Memphis. A year later, he built three more. By the end of the twentieth century, the chain boasted more than 1,100 hotels, and at the time of this writing, Holiday Inns still have ice machines on every floor.

The car, the cooler, packaged ice, and the Holiday Inn together revolutionized once again how Americans use ice. They made ice cheaper and more ubiquitous, but at a cost. Not all ice was being made with the same degree of care or consideration for human health. In some places, especially local grocery stores and gas stations without corporate oversight, ice was made and scooped into bags in back rooms where the possibility of bacterial contamination was high.

In the absence of state and federal health and safety standards for the ice industry, the International Packaged Ice Association created its own standards in 1998. "To be a member of the association, you have to be accredited," said Maria Maggio, executive director of IPIA, in an interview with me. "And the only way to be accredited is by following the association's Packaged Ice Quality Control Standards for sanitation and other requirements, which include having your product tested on a regular basis."

IPIA's regulatory consultant Chris Lamond further explained why having these standards is so important. "If you want to start an ice business today," he told me, "the barrier to entry is pretty low. Most everyone now has a freezer in their house that makes ice that they can put in a bag and sell. If you got a commercial ice-making machine and hooked it up to a hose in your garage—which happens more than we'd care to admit—and started selling ice to the public, well, that food product wouldn't have any federal guidelines that would require you to make sure you're protecting consumers."

In 2005, the Food and Drug Association determined that ice should be subject to the same health and safety regulations it imposes on food products. But given that ice can be made almost anywhere, those regulations are difficult to enforce. According to its website, the FDA "does not inspect small packaged ice producers, like retail stores, that make and package ice directly for the consumer and only for intrastate sales. FDA also does not inspect food service establishments that make ice for direct use."

Contamination outbreaks pose a danger not only to consumers but also to ice companies. "If contamination were to occur from a bag of ice," Lamond told me, "it could potentially have a serious effect on the industry. The ice industry today is primarily made up of large manufacturers. There's little brand recognition among companies, so people might not distinguish between them. If one company's ice made someone sick, people might stop buying ice altogether, at least for a while."

A sudden drop in consumer confidence would be especially detrimental for small businesses like P. Diana and Sons Ice House, which still operates out of its original 1911 building in New Haven, Connecticut. The Ice House originally provided ice for railroad companies shipping perishables from Connecticut to the Midwest. The train tracks, now overgrown and no longer used, can still be seen from the front door. Gary Donaldson runs the business today, following in the footsteps of his iceman father, who took over the company in the 1950s. The family has witnessed firsthand how the ice industry transformed itself over the last century. "My father had the vision that people would buy packaged ice," he told me, "which was something no one really thought would happen. In the fifties, no one thought they'd need to spend money on ice, because we all had ice cube trays. But he was a visionary and knew otherwise."

Today, the Ice House sells packaged ice and block ice. The latter is cut to various sizes and shapes for local bartenders looking to add a wow factor to their cocktails. Sometimes customers stop by to ask for blocks to be cut to fit their personal coolers. As for packaged ice, the company delivers to convenience stores and restaurants. "I've served Bill Clinton and George Bush," Donaldson told me. "Well, not directly, but the restaurants they were eating in ran out of ice and called me. Restaurants always underestimate how much ice they're going to need."

There aren't many businesses like Donaldson's left. National and international ice companies with large automated plants that package hundreds of tons of ice per day now dominate the market. But

Donaldson is making it work. In summer, when demand for ice is highest, he employs a fleet of modern-day icemen: college students on summer break, who deliver bags and blocks of ice to bars, restaurants, weddings, fairs, and festivals all over Connecticut. "We're old-fashioned," he told me with a hint of wistfulness in his voice, "but everyone needs ice."

PART 2

FOOD AND DRINK

CHAPTER 5

The Inside Scoop

Cuban ice-cream peddlers (c. 1900)

I arrived at Mount Vernon during an unseasonably cold spring. A short, no-nonsense guide greeted my tour group wearing a wide-brimmed hat, a lightweight jacket, and an air of intimidating efficiency. She clapped her hands and asked the lot of us, about fifteen or so, to line up. She gave a brief introduction and instructions on what we were and were not allowed to touch inside George Washington's historic home (touch nothing!), and we were on our way.

The tour started with the grounds, which cover five hundred acres in northern Virginia next to the Potomac River. The guide spoke too quickly for me to catch all the details, and my attention wandered. I glanced toward the water and noticed a brick structure the size of a modern carport lodged into the side of the riverbank. I raised my hand and asked what it was. The guide, irritated by the interruption, answered me anyway. It was the estate's original icehouse.

With stairs that circle nearly ten feet into the cold ground, she said, the structure could hold several tons of ice. Someone in the group gasped. To our twenty-first-century ears, that sounded like an outrageous amount. Imagine the work to harvest it: Before Frederic Tudor revolutionized the mechanical aspects of ice harvesting, Washington's enslaved servants would walk out on the frozen Potomac, chisel large blocks of ice with hand tools, and haul them back to shore. The harder the ice, the longer it lasts, so the ice was harvested during the coldest days of the year, when temperatures dropped dangerously low. The actual harvesting wasn't even the most dangerous part of the job. The icehouse was supported with wooden joists that over time rotted from the meltwater. Washington once asked a farm manager to test the joists because "they may give way and destroy those who may be below pounding the ice as it is thrown in."

"What was all that ice used for?" I asked.

The guide explained that most was used to preserve meat and vegetables, but a good amount was set aside every year for making ice cream. America's first president, we learned, was obsessed with ice cream.

A tourist in the back: "I can relate to that."

Mount Vernon's records show that the Washingtons owned twelve "ice plates" and thirty-six "ice pots," which were small cups made specifically for eating ice cream. As the presidential historian Mary V. Thompson has pointed out, "The large number of ice-cream pots suggests that this was a favorite dessert at Mount Vernon, as well as in the capital."

In eighteenth-century America, ice cream was something enjoyed only by wealthy people, because first and foremost, ice cream required ice, and most families did not have icehouses. And as for milk and cream, those were luxuries to be sold at market and rarely kept at home unless money wasn't an issue. Sugar and salt were expensive imported commodities. And the labor required to make ice cream was extensive. Most families couldn't afford to waste their time on such frivolous tasks.

The American Revolution may have severed the country's ties with royalty, but George was still eating like a king.

The Washingtons were likely introduced to ice cream by Norborne Berkeley, 4th Baron Botetourt, who served as governor of Virginia between 1768 and 1770. In May 1784, Mount Vernon acquired its own "Cream Machine for Ice," purchased for approximately one British pound. We also know that George and Martha popularized the dessert among their well-to-do friends by serving it at formal dinners. Each place setting included a pewter "ice" bowl in which the cold treat was served with a spoon.

Beyond these basic facts, tales about George Washington's love for the sweet treat are akin to those about his teeth, his talent for skipping stones across the Potomac, and his inability to tell a lie. They're myths. As one story goes, Martha would put a bowl of cream and sugar on the back stoop every night so that when temperatures dropped, the mixture would congeal into ice cream and George would enjoy it the next day. This is of course empirically untrue—it's impossible to make ice cream in this way. But the tale's persistence and absurdity, writes the historian Marilyn Powell, reveal something of the "essence of ice cream." It has captivated humans for centuries. It has given rise to more myths and legends than almost any other food, except maybe ice itself.

The historian Jeri Quinzio writes that "people spin all sorts of tales about ice cream's origins, and most of them are wrong." Some believe that Nero invented ice cream. He did not. Another origin myth is that Marco Polo brought recipes for "ices" (an early term for ice cream) to Italy from China. Also wrong. "If he had," Quinzio writes, "there would be references to them in the books, letters, and diaries of the time." There are none.

In truth, ice cream probably originated from a lot of different places over the course of several centuries before it evolved into the creamy dessert we crave today. In the Middle East, chilled drinks known as sherbets have been ubiquitous for almost a millennium. Medieval England enjoyed a dessert called "cream of almonde," involving cold milk and nuts. Italian chefs froze cream with saffron to create *crema della mia nonna*, or "my grandmother's cream." Antonio Latini was probably the first to write about making "ices." In 1692, he published the first of a two-volume work called *The Modern Steward*, in which he wrote passionately about the need to balance sugar and liquid to create the perfect "ice." Too much sugar, and the dessert never freezes completely. Too little, and the ice will be too thick to eat.

Not until the end of the seventeenth century, well after the Revolutionary War, did ice cream become common in the New World, and then only in northern states and territories where ice was plentiful. For the privileged few who got to taste it, ice cream was so novel it was something to literally write home about. When a man named William Black ate dinner with the governor of Maryland in 1744, he described the fancy meal in a letter as "strange and decadent" and the ice cream served at the end as "a Dessert no less Curious."

Washington wasn't the only founding president with a sweet tooth for ice cream. Thomas Jefferson kept an icehouse at Monticello, which could hold up to sixty-two wagonloads of ice, a good amount of which he reserved for making ice cream. He first encountered the treat while traveling through Italy. In Rozzano, he tasted several kinds of frozen desserts, noting that "snow gives the most delicate flavor to creams, but ice is the most powerful congealer and lasts longer." Upon returning to the States, he wrote the first recipe for ice cream we know of by an American. His famous recipe called for a splash of vanilla, another rarity in the young country at the time, and included instructions for layering ice and salt in such a way to ensure the mixture reached the desired creamy consistency.

Dolley Madison, wife of President James Madison, further popularized ice cream by holding evening gatherings for any citizen who wished to meet the First Family. While a military band played, guests treated themselves to dishes of ice cream. The wife of a North Carolina newspaper publisher named Josephine Seaton attended one of these parties. The southern woman wasn't impressed, writing, "Pastry and pudding going out of date and wine and ice-creams coming in does not suit my taste, and I confess to preferring Raleigh hospitality." It's possible the ice cream just wasn't up to par. Mary Boardman Crowninshield, the wife of Madison's secretary of the navy, once attended a gathering and described the dessert spread as "ice-cream, put in a silver dish, and a large cake—not good."

Lots of people liked the stuff, however. In 1824, not long after Madison's term ended, a writer named Mary Randolph published *The Virginia House-Wife,* the first American cookbook to have a section dedicated to ice cream. Randolph was among the first to recommend that ice-cream making be moved outside the icehouse. "It is the practice with some indolent cooks," she wrote, "to set the freezer, containing the cream, in a tub with ice and salt, and put it in the icehouse; it will certainly freeze there, but not until the watery particles have subsided, and by the separation destroyed the cream." She recommended placing the mixture instead in a deep pot, which would then be kept inside a large tub surrounded by five inches of salt and ice. Her cookbook sold more copies than any other in the nineteenth century, though some of its purported treats, such as "oyster cream," have not endured.

Randolph was a distant cousin of the Jeffersons. Her wealth and social status gained her access to all the ice she needed for experimenting with ice-cream making. Her readers were likewise wealthy and white. Not until the end of the century, when ice became more affordable and ice-cream-making technologies more advanced, could working people (including white and free Black Americans) make ice cream at home. Until then, they enjoyed it at confectioneries, ice-cream parlors, and elaborately decorated outdoor spaces for music and dancing called pleasure gardens (sometimes called ice-cream gardens). Historically, these places had catered only to the elite, but in the face of so much national change—more class mobility, more racial consciousness—many were making changes of their own, including welcoming people of all economic backgrounds. Most remained racially segregated, but those that were integrated were owned by Black chefs whose groundbreaking recipes elevated ice-cream making to something like an art. Indeed, the oft-told history of American ice cream has tended to focus on presidents and their inner circles, but in truth, it was the most marginalized of Americans, the poor and racialized classes, that turned it into the nation's favorite dessert.

———

To make ice cream, one needs ice, so it comes as no surprise that the first American shops to sell ice cream were located in early nineteenth-century Philadelphia, Boston, and New York, cities where ice harvesting was common and icehouses were already built. Eleanor Parkinson opened one of the nation's first confectionery shops in Philadelphia in 1818 in a small room adjacent to her husband's tavern. She sold cakes, chocolate, bonbons, and elaborate ice creams made with an inordinate amount of sugar—each batch of her ice cream contained three cups, while the most common recipes of the day called for just one.

Little is known about who visited her shop, but it was likely segregated. Most confectioneries were then. That's one reason why Augustus Jackson, a free Black man who had worked in the White House kitchen during the Madison administration, opened his own confectionery in the neighborhood now called the Center City District. Knowing that most Americans, regardless of race, preferred "pure" ice (that is, ice harvested from rural lakes rather than polluted rivers flowing through cities), he purchased the purest he could afford. He experimented with fruits and nuts to create unusual flavors, and he used fresh cream solids, which, when frozen, created a tantalizingly thick mixture that coated the tongue—a far cry from the watery substance served on the Washingtons' table.

Jackson's ice creams created a sensation. Other Black cooks followed his lead, opening confectioneries of their own in Philadelphia. Two of them asked Jackson to supply their shops with his frozen delights, and he did, essentially launching with his ice cream one of the country's first franchises. Jackson's shop catered to people of all races and economic backgrounds, and his ice cream was, for most patrons, the first they'd ever tasted. Jackson would die a wealthy man and one of only a handful of Black men in the city who owned his own icehouse.

Meanwhile, in New York, pleasure gardens had become all the rage. Modeled after large outdoor courtyards in Europe, New York's "gardens" were small patches of green, often adjacent to taverns, where patrons could sit in the sun and enjoy a drink and lemony "ice." Much of Manhattan was still rural then, and its first pleasure garden, the Vauxhall Garden, was considered to be too rustic to attract metropolitan customers when it opened near Astor Place in 1805. This was one year before Frederic "the Ice King" Tudor launched the ice trade down in the warmer southern climes and ice cream was still a novelty, even for New Yorkers. They flocked to the Garden.

Over the next twenty years, the city's pleasure gardens evolved into performance spaces with colorful lanterns, theatrical presentations, sculpture exhibitions, and fireworks. Glasses of lemonade were served over sparkling slivers of ice—something rarely seen in restaurants. Most gardens claimed not to serve liquor, but, according to Abram C. Dayton's *Last Days of Knickerbocker Life in New York,* "a quarter slyly dropped into a sable palm would ensure a moderate supply of cognac to be poured over the lemon ice." The gardens' popularity peaked in the late 1850s. It was then that, at the age of seventy-five, Frederic Tudor himself opened a pleasure garden just north of Boston called Maolis. His establishment included a teahouse, a dance hall, a "Witch House" for children (akin to a haunted house today), and of course an ice-cream pavilion, whose ice was likely supplied by Tudor's own ice company.

The democratization of pleasure gardens did not sit well with New York's upper classes, who by now had mostly abandoned them, save for a few on Broadway. The theater district's more expensive gardens, writes the historian Naomi J. Stubbs, strived "to define and establish themselves as being elite and distinct" from lower-class establishments by keeping the classes apart. They succeeded by hiking their price of entry far above what a working-class person could typically afford. A handful of newspapers helped cement the public's opinion of pleasure gardens as rowdy, dangerous places by publishing reports of drunkenness and riots. In actuality, most pleasure gardens rarely witnessed such disruptive behavior.

This shift in economic class wasn't the only change afoot at the gardens. Like white-owned confectioneries, white-owned pleasure gardens rarely welcomed Black patrons (though some hired Black servers). As one news report put it, "Among the number of ice cream gardens in this city, there was none in which the sable race could find admission and refreshment." That changed in 1821 when a Black entrepreneur named William Alexander Brown opened a pleasure garden in lower Manhattan. A local newspaper called the establishment the "African Grove," a place where "ebony lads and lasses could obtain ice cream, ice punch, and hear music from the big drum and the clarionet." But the garden didn't last long. Less than a month after opening, Brown received complaints from white neighbors that forced him to close.

Brown had a talent for entertaining, however, and a drive to do it. According to the historian Marvin McAllister, Brown went on to open one of the country's first racially integrated theaters. Alas, assaults by racists forced him to close that business, too. On his theater's last day of operation, he put up a sign that read: "Whites Do Not Know How to Behave at Entertainments Designed for Ladies and Gentlemen of Colour."

Back in Philadelphia, a Black refugee from Saint-Domingue (now Haiti) named Monsieur Collot opened an integrated pleasure garden that served Italian-style ices. Having left his home country during the slave rebellions, Collot was most familiar with Western European ice-cream recipes brought to the West Indies in the early eighteenth century. He boiled his ice-cream mixture of eggs, cream, and sugar until it resembled custard, and then cooled it for several hours in a tub of splintered ice sprinkled with salt. Like Jackson's recipes, Collot's called for the purest ice available, though little of it actually came into contact with the mixture's edible parts.

Jackson and Collot inspired other Black entrepreneurs to open confectioneries and pleasure gardens. A free woman named Sallie Shadd opened a shop in Wilmington, Delaware. Some historians argue it was her recipe that Dolley Madison served in the White House. In New Jersey, Peter Scudder, a shoe shiner and apple peddler, was one of the first Black men to sell ice cream in what's now Princeton. In Kansas, two Black women—Miss Ingrams and Miss Holmes (their first names are lost to history)—opened one of the state's first confection-eries in Atchison. In Kentucky, Mrs. Susan Green became one of the first ice-cream vendors in Lexington and the city's only Black ice-cream vendor. And in Erie, Pennsylvania, a former ice-cream factory worker named John S. Hicks opened his own integrated confectionery—and then an ice-cream factory—that allowed him to sell ice cream at the lowest price in the state. By the end of the nine-teenth century, Black-owned ice-cream shops were open across the country, attracting not just Black patrons, who could go nowhere else to purchase ice cream, but also working white people who could only afford to visit these Black-owned shops. Ice cream had finally become a dessert for the masses.

In 1897, a Black man named Alfred L. Cralle invented the ice-cream scoop. Though patented, his invention was quickly and freely adopted by places everywhere selling ice cream, and he never made a profit from it.

New York City, 1855. An ice deliveryman hoisted a 50-pound block of ice from the back of his wagon and heaved it over his shoulder. His other hand gripped a pair of metal tongs, which held another 50-pound block. As he carried the ice to his client's front door, the block dangling from the tongs slipped loose and fell to the ground. He swore loudly, but no one heard him, because just as the ice hit the ground, an ice-cream peddler in a tattered shirt and thick wool pants rounded the corner. The peddler's voice was so loud it pierced through the sounds of shouting men, boys hawking newspapers, and hundreds of horse hooves clopping down the street. The peddler had to be loud. That's the only way the children up on the third and fourth stories could hear that he'd arrived.

The ice-cream peddler was a familiar figure by the mid-nineteenth century. By then, the ice trade had brought down the cost of ice enough that in large cities even the poorest residents could afford small amounts at least some weeks of the year. At the same time, ice cream was becoming a favorite treat among people of all races and economic classes.

Black Americans in particular who couldn't afford to open their own confectioneries saw in these converging trends a lucrative opportunity: they could sell ice cream in the streets. Some carried tins of ice cream on their shoulders without any ice to keep it cool; the ice cream would melt quickly. Others pushed wooden carts filled with cheap (and often polluted) ice that kept the tins cool for at least a couple of hours each day. The unsanitary conditions in which they served ice cream would turn stomachs today. They carried small glasses called penny licks that they'd fill with ice cream and hand to customers. The customers would either lick the dessert directly from the glass or wipe it out with their fingers. When they returned the empty glass, the vendor would swirl it around in a bucket of warm, gray water before refilling it for another customer.

Every peddler sold his wares by shouting in the streets, a sound that earned mixed reactions from neighbors. Children loved the peddlers, of course. They could buy an ice cream from them for merely a cent. But confectioners and pleasure-garden owners resented what they saw as a threat to their businesses.

Some social reformers also disdained the peddlers because the peddlers' customers were mostly poor. An article in the *New York Herald* described impoverished children crowding around an ice-cream cart: "From morning until night the children stuff themselves. . . . Where does the money come from? . . . Thriftless, but affectionate, is the lower class parent. Shoes the child must do without, for the father has not quite enough money to purchase them." Newspaper reporters often dismissed the peddlers out of hand, using the men's race or ethnicity as reason enough to avoid them. One such critic boasted that while he had never tasted a Black peddlers' ice cream, he was sure that the "African article will not bear a comparison with Parkinson's," a reference to Eleanor Parkinson's white-owned confectionery.

In the second half of the nineteenth century, Black ice-cream peddlers were joined by Italian immigrants who had come to America to escape the independence wars of their home country. Most arrived without a trade and could not speak English. American politicians called them "good-for-nothing mongrels" and "human flotsam." Most of the immigrants were controlled by *padron*es, Italian bosses who had immigrated earlier, could speak good English, and knew how to leverage their experience in America to dominate and profit off the new arrivals. The bosses provided housing for the newcomers in crowded tenements, where cholera ran rampant. They took advantage of the new immigrants' poor writing and reading skills by tricking them into labor arrangements that paid barely enough for them to eat. Many immigrants became stevedores or ice harvesters. Some became organ grinders—a street job that required no musical talent, just the ability to turn a hand crank. Many, though, became ice-cream peddlers, who expanded the trade throughout New York into every neighborhood.

———————

A scientific discovery at the end of the nineteenth century almost brought an end to ice-cream peddling. Germ theory was still little understood, but newspapers reported how scientists studying natural ice were finding "bugs" and "smells" that were making people sick. Natural ice was the only kind that peddlers used, because it was what they could afford. Thus, Americans—especially those with the option to visit parlors—became wary of ice cream sold on the street. When a study revealed dangerous levels of bacteria in the peddlers' ice cream and bucket water, Americans' confidence dropped further.

The bosses who owned the carts that immigrants rented consolidated their businesses, and cities set regulations on what these businesses could and could not do. By the late 1890s, ice-cream vending had become as regulated as the ice industry. New York passed laws that forbade the making of ice cream in tenements, regardless of whether it was for personal use or selling. (This law was likely influenced by the lawmakers' general suspicion of tenements, which they associated with immigrants, and therefore filthy conditions.)

Peddlers were forced to buy their ice cream directly from wholesalers, who were becoming more numerous by the day. And to operate a cart, vendors were required to purchase a ten-dollar license—a hefty price for people who made just a few dollars each month.

At the turn of the twentieth century, another change rocked the ice-cream business. As with the ice industry, the ice-cream industry was about to explode because of a trailblazing marketing campaign. In 1902, Harry Burt of Youngstown, Ohio, began delivering ice cream from a motorized vehicle, when most other people were still relying on wagons pulled by horses. The motor enabled him to stay out for longer periods of time and bring his treats to more customers. By 1920, he owned a fleet of these newfangled "ice-cream trucks" and inspiration struck again. He began selling small bricks of ice creams coated in chocolate and stuck with lollipop sticks. He called the treat a Good Humor ice-cream bar, and to sell it, he painted his trucks white and made his men dress in crisp white suits. Their image suggested wholesomeness and cleanliness—a strong departure from the "dirty" peddler or the "salacious" iceman—the kind of men that husbands could trust to be around their wives and children. He decked the trucks with bells taken from the family bobsled, and as they drove through town, the bells beckoned the children just as the peddlers' voices once did.

The Good Humor Ice Cream Company eventually switched from ice buckets to refrigerated trucks, but the image of the clean and wholesome Good Humor man changed little through the first half of the twentieth century. An idealized Hollywood version of the character appeared in dozens of movies in those years, including *The Good Humor Man* of 1950, starring Jack Carson in the title role. By then, Good Humor marketing was so all-encompassing that someone, probably an ad executive, spread a rumor that the head of Columbia movie studios stopped a heated argument with his brother during filming when he heard an ice-cream truck's bell on the street. As the (possibly apocryphal) story goes, he left the set, ran to the truck, and ordered three Good Humor bars before returning to deck his brother in the jaw. Now that's good ice cream.

As ice cream became more and more a feature of everyday life, another frozen treat was catching on. On a cold night in 1905, eleven-year-old Frank Epperson left a stir stick in a glass of water mixed with flavored drink powder out on his back porch. The next morning, the flavored water had turned to ice in the glass, which he ate while holding the stick. In 1923, a grown-up Epperson founded Epsicles, later called Popsicles, and sold them wherever people gathered for fun: at amusement parks and beaches, mostly. A stand at Coney Island reportedly sold eight thousand in a single summer day. A year later, Epperson's company sold more than 6.5 million of his icy "drinks on a stick."

In the wake of so much change in the ice-cream industry, the days of the elite confectionery were coming to an end. The ice-cream soda fountain bubbled up in its place, serving carbonated beverages mixed with scoops of ice cream. Customers loved the icy, fizzy mix, especially teenagers. In an 1891 edition of *Harper's Weekly*, Mary Gay Humphreys wrote, "On a bright exhilarating day, to achieve a cup of ice-cream soda, a place should be engaged some time in advance. Beauty and fashion surge about the counter. One of the sights of the town is the rows of bright faces, two and three deep, bent over their cups, and fishing within with long-handled spoons." In *Ice Cream: A Global History*, Laura B. Weiss writes that sodas "were capable of inspiring near rhapsodic reactions." An Omaha newspaper reporter wrote in the 1880s that an ice-cream soda is "a drink that combines one ice cream and one soda in a moonlight sonata of perfect harmony." By the 1910s, over 475 million gallons of soda water and at least as much ice were consumed at American soda fountains every year.

During Prohibition, the popularity of soda fountains vaulted to new heights. One New York pharmacy reported selling 100 gallons of ice cream and 9,000 pounds of ice in a single day. Saloons closed their doors and reinvented themselves as soda fountains. The St. Louis brewer Anheuser-Busch, which owed its existence to the ice trade, made use of its ice cellars by making and selling ice cream. By 1929, 60 percent of the country's 58,258 drugstores had installed soda fountains. They could also be found on trains and in airports, tobacco shops, and department stores.

The popularity of these places was buoyed by another innovative frozen dessert—the sundae. As with ice cream, no one knows for sure where the sundae came from, but one origin story begins in Ithaca, New York, where a Unitarian minister named John M. Scott visited a drugstore after services one Sunday and ordered a dish of vanilla ice cream. Chester Platt, who operated the soda fountain, topped the reverend's dish with cherry syrup and a candied cherry. The dish tasted—and looked—so good that Platt kept it on the menu. An ad in the *Ithaca Daily Journal* crowed about the drugstore's revelatory "Cherry Sunday," a "new 10 cent Ice Cream Specialty."

In college towns across the country, ice-cream Sundays (eventually "sundaes") were called "college ices" and were said to be especially popular among female students. During Prohibition, many beer-drinking men who had grown accustomed to spending their evenings in boozy saloons found it hard to adapt to the clean-cut environs of the soda fountain. Only "little girls and dudes drink ice cream and soda," scoffed one *Chicago Tribune* reporter. Sensing men's hesitancy to partake, druggists set aside separate areas where women could enjoy their sodas and ice cream apart from the chagrined men who drank only soda. Fountains located in department stores catered especially to female customers by calling their specially blended flavors the "Queen's Favorite."

While some soda fountains were segregating the sexes, others were working to integrate the races. On June 23, 1957, three years before the Greensboro sit-ins, the Reverend Douglas Moore led six Black students into the segregated Royal Ice Cream Parlor in Durham, North Carolina. The protesters walked through the parlor's "colored" entrance in the back, then proceeded to the white section, where they sat down and ordered ice cream. The soda jerk refused to serve them. The manager told them to leave. They responded by trying to order again. The group was arrested and fined ten dollars each. Years later, one of the young protesters, Virginia Williams, said this of her experience: "If he had served us ice cream, he would have made history. But, by refusing to, I guess we made history!"

A s demand for ice cream grew, so did demand for ice. In his 1919 *Book of Ice-Cream*, Walter W. Fisk estimated that ice-cream companies needed 614 pounds of ice per day to churn. Hardening and storing ice cream demanded another 914 pounds, and to ship and deliver ice cream, a company needed nearly 400 pounds of ice.

By the early 1920s, the cost of commercial ice was more than the cost of electric refrigeration, so the ice-cream industry switched over. The industry changed again, when a decade later, the Rural Electrification Act brought electric refrigerators to homes across the country. After World War II, the California-based chain Baskin-Robbins introduced the notion that ice cream could come in more than vanilla, chocolate, or strawberry by offering an outrageous thirty-one flavors, one for every day of the month. This innovation paved the way for the imaginative flavors we see today, flavors such as the San Francisco–based Humphry Slocombe's bourbon-and-cornflakes-infused Secret Breakfast. A scoop of that once pinged the pleasure center of my brain so hard it brought tears to my eyes.

In 1959, Reuben Mattus, who got his start in the ice-cream business by vending his mother's lemon ice in the Bronx, started Häagen-Dazs. By the early 1970s, Häagen-Dazs was distributed by every major grocery chain in the country. Multinational companies such as General Mills, Nestlé, and Unilever launched their own brands of ice cream, some with healthy twists on the original recipe. Low-sugar, sugar-free, low-carb, and even dairy-free versions of ice cream began filling supermarket freezers.

Today, ice cream makes for sweeter birthday parties, happier weddings, and easier breakups with romantic partners, but it's hard on the planet. The production of a single pint of ice cream spews between 2 and 4 pounds of carbon dioxide into the atmosphere, or approximately the same amount the average gasoline-powered car produces on a two-mile drive. That CO_2 comes mostly from dairy cows, which produce the greenhouse gas methane from both their front and rear ends. Transporting ice cream from the factory to distribution centers spews even more CO_2, as does all the packing and eventual freezing of the product.

This is all very bad for the environment, but some ice-cream companies are taking action to reduce their impacts on the planet.

The Vermont-based brand Ben & Jerry's has been one of the industry's most forthright companies about ice cream's environmental toll. "Our world is already seeing the devastating effects of climate change," reads the company's website, "and time is running out to act to avoid even more catastrophic consequences." For its part, the company has set goals to use only renewable energy sources by 2025 and to cut its greenhouse gas emissions by 40 percent.

Unilever, one of the planet's biggest commercial food companies and owner of Ben & Jerry's, is seeking to reduce its carbon footprint by changing how it refrigerates ice cream. In 2021, the company launched a pilot program in the Netherlands in which it replaced four diesel-fueled refrigerator trailers with zero-emission battery-electric prototypes. "If successful," the company's website reads, this change "could save up to 25 tonnes of CO_2 per trailer annually, with air quality benefits for each vehicle equivalent to taking 70 passenger cars off the road for a year."

Despite its environmental impacts, ice cream isn't going anywhere. In 2021, U.S. commercial ice-cream makers churned out over

1.3 billion gallons of ice cream, made $13.1 billion, and supported over 28,800 jobs. The historian Anne Cooper Funderburg writes that ice cream has become so ubiquitous, so profitable, and so mythically large in the nation's imagination that it is "a symbol of the American dream." George Washington probably would have agreed.

CHAPTER 6

"Culture in a Glass"

An iceman waits for a phone call for delivery (1915).

While exploring the archives at George Mason University in Fairfax County, Virginia, I took a break to visit the university's dining hall for a hit of caffeine. It was a hot day and I wanted iced tea, but here I was south of the Mason-Dixon Line, and the only options came presweetened. Sugared-up tea is not, well, my cup of tea. As a northerner, I prefer my tea black.

Sweetened iced tea, or "sweet tea," as it's often called in the American South, has become emblematic of southern culture and tends to be the default iced-tea option in restaurants and dining halls throughout the region. Food writer Allison Glock wrote in *Garden & Gun* that sweet tea "isn't a drink, really. It's culture in a glass." In *Holy Smoke: The Big Book of North Carolina Barbecue,* food historian John T. Edge is quoted as saying that sweet tea is "a kind of culinary-cultural Global Positioning System, an indicator of where we are and, yes, who we are." The southern cookbook author and North Carolina chef Sheri Castle describes sweet tea as "a rite of passage. Being offered a glass of sweet tea instead of milk is how Southern children know that they're growing up." It's so emblematic of the South that southern-themed restaurants in northern cities, such as Chicago, Seattle, and New York City, all default to ice-cold sweet tea on the menu.

The beverage also features prominently in popular culture celebrating the South. In his love letter to the region, a song called "Chicken Fried," the Georgian country artist Zac Brown lists things that he considers inextricable from southern life: "Sweet tea, pecan pie, and homemade wine."

Blake Shelton sings about love and sweet tea in "Doin' What She Likes" ("little kisses, sweeter than sweet tea"). So does Scotty Mc-Creery in "Southern Belle" ("If lovin' those sweet tea, blue jean dreams is wrong / Well, I can't help it"). And sweet tea is more than a randy metaphor in Jake Owen's "Barefoot Blue Jean Night"; it's literally what his love interest is drinking ("Her ruby red lips was sippin' on sweet tea"). Songwriters have for decades been getting hot and bothered over iced tea made sweet.

They've been getting drunk on it for even longer. Some of the earliest known recipes for sweet tea called for an inordinate amount of alcohol. Lettice Bryan's 1839 cookbook, *The Kentucky Housewife*, includes a recipe for "tea punch" that involves pouring hot tea over sugar before mixing in cream and several cups of chilled champagne or claret wine, then serving it over ice. The southern military was famous for the strength of its "tea punches." Chatham Artillery Punch, the signature drink of an elite Savannah militia unit, called for mixing massive amounts of tea and citrus to hide the recipe's dizzying amount of alcohol.

The nonalcoholic version of sweet tea that we know today first appeared in the decade before the Civil War. In 1856, the Richmond, Virginia, druggist S. P. Semple advertised that at his soda fountain, "the exhilarating effects of a glass of iced Tea or Coffee [would] speak for themselves."

Sweet tea has been drunk in the region for almost two centuries, but despite today's overwhelming association of sweet tea with the South, the drink didn't catch on there until the second half of the twentieth century. That's because "sweet tea" is a form of "iced tea," which requires plentiful amounts of ice to serve properly. Even after the launch of the ice trade, which brought northern ice to the southern territories as early as the 1820s, ice in the South remained an expensive luxury. Not until the establishment of affordable home refrigerators and the rise of car culture did ice make its way out of wealthy neighborhoods in port cities and into poorer neighborhoods and more rural communities. How, then, did it come to be thought of as the iconic southern drink we know it as today?

Like the history of ice, the history of iced tea in the United States is subject to much debate. But most historians point to the 1850s as the leafy drink's breakout decade, which began, surprisingly, in the North. For the previous fifty years, the country had been experiencing a rise in alcohol consumption, leading to all kinds of health and social problems. A growing faction of Christians began calling for temperance, and their cries gained traction in national politics. By 1855, thirteen states had outlawed hard liquor, but all this did was reduce the rate at which Americans drank spirits; they still loved their beer.

What the temperance movement needed to ween a nation off booze completely was a tasty alternative, something that refreshed while still providing hints of both sweetness and stringency. The solution, thought some of the movement's leaders, was iced tea. One Christian group ran an ad in an 1857 edition of the *Saturday Evening Post* in an attempt to persuade the public to take up tea drinking: "Tea made strong," it read, "well sweetened, with good milk or better cream in it in sufficient quantity to give it a dark yellow color, and the whole mixture cooled in an ice chest . . . is the most delicious, the most soothing, the most thirst allaying drink." The ad did little to convince Americans to adopt iced tea as their beverage of choice. Just two months later, the *Post* published an editorial by a writer disheartened by the whole situation: "Our saloon keepers don't advertise these delightful drinks 'which cheer but not inebriate.' . . . . We suppose it will be a century before the public finds out what luxuries iced tea and coffee are in the summer solstice."

They actually didn't have to wait that long. In July 1868, the *Boston Journal* declared, "During the heated term there is nothing so invigorating as iced tea. A slice of lemon no thicker than a wafer placed in each tumbler adds to the relish." The *New-York Commercial Advertiser* ran the same notice five days later. The *Chicago Daily Inter Ocean* commented, "Iced tea with a slice of lemon in it is said to be decidedly ahead of lager," while the *Harrisburg Telegraph* called iced tea a "summer exhilarator."

During the next two decades, iced tea would become, if not quite a staple of northern gastronomy, then something quite common in the region. In 1889, Sarah Tyson Rorer wrote the following in her popular newspaper column Table Talk: "Twenty years ago the fondness for that beverage was confined to a few who were looked upon as 'gastronomic cranks.' Today, we are rather inclined to think there is something cranky about a man who says he doesn't like iced tea."

The beverage continued to steadily gain in popularity in the North. Then, at the turn of the twentieth century, its popularity exploded practically overnight in the Midwest and, eventually, in the South. The reasons that iced tea finally caught on are many, but one of the biggest contributing factors was a not-so-little get-together planned in St. Louis, Missouri, in 1904.

When folks today think of St. Louis, ice isn't usually the first thing to come to mind—beer is. The city is still home to the Anheuser-Busch brewery, which at the time of this writing is the largest brewer in the United States. The Lemp Brewery, the first to brew lager in America, also got its start in the Gateway City. In the decades before mechanical-ice production, both breweries purchased blocks of natural ice from the small town of DePue, Illinois, located about two hundred miles up the Illinois River on a swath of flat prairie largely untouched by the Industrial Revolution. Its surrounding lakes were known for their clarity. An Englishman named Charles Stedman owned most of the area's ice industry. He signed an exclusive contract with Lemp that called for the shipping of 50,000 tons of ice harvested by more than three hundred men every year, and at least fourteen steamboats to carry the ice to the city.

In the 1870s, Lemp supplemented these shipments by building its own mechanical-ice plant that produced "150 tons of artificial ice" every day. Between the ice plant and the brewery's weekly ice shipments, ice became something of a point of pride for the people of St. Louis. The city was now the fourth largest in the country, and ice had become as much a way of life there as it had in New York, Chicago, and Philadelphia. Ice companies that harvested blocks from the Illinois River made deliveries to homes throughout the city and to farms located just outside the city limits. St. Louis bartenders were among the first (just after those in New Orleans) to serve cocktails shaken over ice. An icy cocktail mixed with rum and lime juice called planter's punch became the hottest (and coldest) drink in town. A Tom Collins served over ice also became a favorite drink among St. Louis bartenders, though experts debate whether the drink was actually invented there.

Then, in 1904, ice took on a mythic quality in the city. That year witnessed the opening of the Louisiana Purchase Exposition, also known as the St. Louis World's Fair. It changed forever how the people of St. Louis—and fairgoers from around the world—thought of "cold."

The fair featured exhibits from fifty nations and drew nearly twenty million visitors. It popularized hot dogs and ice-cream cones and inspired Andrew Sterling and Kerry Mills's song "Meet Me in St. Louis." Visitors marveled at technological advances such as X-ray and fax machines, and the first electric typewriter. But the star of the show was the fair's ice plant, which engineers modified to create one of the world's first and largest electric air conditioners.

An official fair photobook notes that a "refrigeration plant installed in the basement [of the Missouri State Building] has the capacity to reduce the temperature in the building to seventy degrees even when the mercury may be in the nineties outside." The plant cost an astronomical $127,000 to build and install, but the effect it had on attendees was nothing short of mind-blowing. For most fair-goers visiting St. Louis in high summer, the drop in temperature would have felt like magic. "Visitors not aware that the building was artificially cooled," said an article in that year's edition of *Ice and Refrigeration*, "were struck with wonder and were unable to account for the very perceptible change felt in the temperatures."

That same article explained that the cooling system operated not unlike central air-conditioning in today's typical American homes: it used ammonia to cool the air and electricity to pump the air into buildings. The technology mirrored that used in the Lemp and Anheuser breweries' ice plants, and for the first time in the history of the world's fair, it was used to create both air-conditioning and a large supply of ice.

So much ice. More than 25 million pounds in eighteen months, making it one of the largest and most productive plants in the country. The ice helped keep the ice cream cool throughout the fair. It also helped the nation's burgeoning flirtation with iced tea to blossom into true love.

Over in the fair's India Pavilion, a British tea plantation owner named Richard Blechynden was promoting black teas of India and Ceylon (now Sri Lanka). That summer, a heat wave gripped the city of St. Louis, sending outside temperatures soaring to ninety degrees or more. Attendance at the fair dwindled as temperatures rose, and those who stayed flocked to either the air-conditioned building or to exhibitors serving ice cream. Blechynden had intended to sell his tea hot—the traditional method in India and Britain—but in such heat, few passersby were willing to stop to try a steamy cup. Multiple stories exist of what happened next. One is that Blechynden somehow utilized the plant's pipes to cool his tea. Another has him borrowing a metal bucket and making his way to the fair's engineering office, where the plant's engineers agreed to fill the bucket with chipped ice. Back in his booth, he would have poured his tea over the ice and sold it to overheated fairgoers. The drink was a sensation.

After the fair closed, Blechynden took his iced tea to New York City, where he persuaded the owners of Bloomingdale's to sell it to shoppers. The beverage spread to other cafés in the city, where out-of-town visitors developed a taste for the drink and brought it back to their hometowns. By the 1920s, iced tea was in cafés both above and below the Mason-Dixon Line. That decade also witnessed the birth of the electric refrigerator, which allowed middle-class families to keep (and in some cases, make) ice at home, further increasing iced tea's popularity. As the historian Joe Gray Taylor writes in *Eating, Drinking, and Visiting in the South: An Informal History*, the electric refrigerator was "probably more appreciated for the ice cubes it provided . . . than for any of its other services," especially in the rural South.

To be clear, Blechynden didn't invent iced tea. As the world's fair historian Pamela J. Vaccaro points out, an exhibitor named N. B.

Reed had earned over $2,000 selling iced tea nearly ten years earlier at the 1893 Chicago World's Fair. (And as we've seen, cookbooks featured recipes for iced tea decades earlier.) The difference between that earlier fair and this one was the invention of photography. The medium, which had been used sparingly to document world's fairs since the first in London in 1851, had by 1904 become a common tool of journalists working for newspapers and magazines across the country. Photos of the St. Louis World's Fair were published everywhere, and they helped spread word of the fair's wonders, including iced tea, to people of all kinds. Another driving force behind iced tea's popularity was the ice itself.

By 1904, restaurants and cafés had adopted commercial-sized iceboxes so that they always had ice on hand to chill drinks and make frozen desserts. Like never before, the gastronomic world was ready for icy innovation at a commercial scale, and iced tea perfectly fit the bill: tea was cheap and readily available, and recipes for iced tea didn't call for anything special beyond ice.

In 1928, Mrs. S. R. Dull published *Southern Cooking*, an influential cookbook with perhaps the best-known iced-tea recipe in the United States, a recipe still used at home by aficionados today: "Steep hot tea until desired strength (which should be strong since it will be poured over ice) and add sugar while the tea is hot to ensure it melts and infuses well." The author recommends garnishing it with lemon, mint, strawberry, cherry, orange, or pineapple. She is unwavering in her opinion that milk should never be added to iced tea.

The popularity of iced tea spread south, where it became a year-round drink. But as Robert Moss points out in an article for *Serious Eats*, the beverage didn't become a matter of cultural identity until, incredibly, at least the 1980s. That decade witnessed the formation of a cultural apparatus dead set on defining—or perhaps redefining—what it meant to be southern. In the 1989 movie *Steel Magnolias,* Dolly Parton's character, Truvy Jones, calls iced tea "the house wine of the South." In his 1987 seminal work of food history, *Southern Food: At Home, on the Road, in History,* John Egerton offers a list of what he calls "classic" southern foods: beans, greens, chicken and dumplings, and, sure enough, iced tea.

In 1993, food writer Jay Grelen wrote an article in the *Mobile Press-Register* explaining that sweet tea was a symbol of the South. (He'd go on to write a column about southern life and culture for the *Arkansas Democratic-Gazette* called simply Sweet Tea.) That same year, a series of polls conducted by the University of North Carolina revealed that "seven out of eight Southerners drink iced tea, and two-thirds of those prefer sweetened." In 1994, a writer for Alabama's *Anniston Star* threw down the gauntlet with "Southern ice tea is always very sweet." In fact, "it's not Southern tea (and in the South isn't worth much) if it doesn't have a lot of sugar added to it way in advance."

Sweet tea was becoming fully metonymic for the South, appearing as a stand-in for the region in popular songs, movies, and a plethora of restaurant and food-company marketing plans. But why? Especially given its original connections to the North and Midwest?

In a 2007 article for *Slate,* Jeffrey Klineman posits that sweet tea's "appeal lies in the ice. Southerners seem to have a particular fascination with ice. This may stem, most obviously, from the fact that the Southern climate is often steamier than a Rat Pack schvitz." It may also have something to do with what the beverage represents—or rather, what it doesn't.

The post–Civil Rights era witnessed a South changing culturally and economically. As Moss has argued, this was a time when white southerners were "searching for things to anchor our collective identity," because previous symbols, such as the Confederate flag, were becoming more widely understood as representative of slavery and Jim Crow. And for obvious reasons, Black southerners were never attached to those symbols to begin with. The word "Dixie," writes the sociologist John Shelton Reed, was "losing some of its currency" as race relations improved slowly and businesses chose to avoid the word. Iced tea, devoid of antebellum connotations, became a stand-in for what some were calling the New South.

At the same time, Moss continues, the suburbs of the South were experiencing the same kind of gastronomic flattening as those in the North; the "fast food chains that lined our suburban highways" in Atlanta were the same as those in Chicago, St. Louis, and Philadelphia. Every downtown had a McDonald's. Every interstate off-ramp led to the same Shell station. Whatever had been unique about southern culture was slowly being replaced by a more generic "American" culture. So, sweet tea became for southerners a point of connection that felt harmless yet distinctive enough to read as special.

It wasn't long, though, before the Americanization of the South absorbed sweet tea, too, and the newly southernized drink crept back north. Snapple and Nestea began selling bottled sweet tea, and in the mid 2000s, McDonald's added sweet tea to its menu. In the 2010s, the southern chain Chick-fil-A opened restaurants in Chicago and New York, ingratiating its version of iced sweet tea into the hearts of northerners. And every summer, America's largest coffee chain, Starbucks, offers at all its cafés iced tea shaken with several pumps of simple syrup. If *Steel Magnolias* were remade today, a fact-checker might push to revise Truvy's most famous line: "As *American* as sweet tea."

CHAPTER 7

On the Rocks

A bartender serving another drink

When I stepped inside the lower Manhattan cocktail bar Weather Up, my eyes took a moment to adjust to the dim light. I couldn't tell whether the shadowy figure sitting in the back was my friend Tana or a woman who just looked like her. The figure waved, so I walked over. Tana stood up and hugged me. It had been too long since we'd seen each other. We spent some time catching up and then gave each other a knowing look.

"Are you ready?" I asked.

"Let's do this." Tana picked up the cocktail menu.

Weather Up's cocktails are legendary. I ordered a whiskey-based one, something akin to an old-fashioned but with more citrus, and she ordered a twist on the gin fizz. When the drinks arrived, we clinked our glasses, then paused to admire the ice before taking our first sips. The cube in mine was short and square, and clear as crystal. Tana's was long and lean, equally clear, and popping in the carbonation.

The drinks tasted like they looked: perfect.

Not that this surprised us. This was Kathryn Weatherup's second bar bearing her name (the first was in Brooklyn), and since opening the place in 2010, she and her team have made headlines for the precision with which they make their cocktails—and for the quality of their ice.

As Weatherup's business partner Richard Boccato told the *New York Times* when Weather Up opened, "We're going to be the first bar on the East Coast of the United States [to do] in-house ice harvesting and production." That work is done in the basement, which houses an ice machine called the Clinebell CB300X2 Carving Block Ice Maker. It's the kind used by ice companies to produce 300-pound blocks for selling to ice sculptors—the clearest ice on the market. The machine cost $6,000 in 2010.

During the first bar's early years, chunks of ice were displayed throughout, having been cut into appealing shapes by "bandsaw, chainsaw, chisels, hammers, and other torture devices," as Boccato put it. When I visited, nearly twelve years after its opening night, the ice displays were gone and the Clinebell was temporarily shut down because of the pandemic. But the ice in the cocktails was still made in-house and cut as with a surgeon's hand. I asked Weatherup why ice was so important that her bartenders made it themselves.

"It really comes down to the basics," she told me. "The lick of the citrus, the modifiers, the glasses you use, and the ice—that's what you need to make a good cocktail." The way a bartender plays with each element, she continued, changes how the cocktail tastes.

I asked Weatherup what inspired her love of ice, and she paused a moment before mentioning Sasha Petraske, the founder of the New York cocktail bar Milk and Honey, who died at the age of forty-two in 2015. "I never worked at Milk and Honey," she said, "but I did all my training there. It was [Petraske's] passion and knowledge of cocktails that started this whole trend [of using clear ice] in bars."

Milk and Honey became one of the most influential cocktail bars of the early twenty-first century, with its pre-Prohibition-era drinks and what the *New York Times* once called "eccentric reservation system and exacting rules of decorum." The bartenders sported formal attire, and the accidental spilling of drinks, whether by the bartenders or the patrons, was frowned upon. This wasn't a place to get sloppy drunk.

In contrast, Weatherup said that her bars were never meant to be "fancy." They were "bar bars," the kind where it was okay to get a little tipsy, even a little loud. To her mind, this took nothing away from the quality of the drinks or the spectacle of the ice. At the end of the day, "there's something quite fascinating about a crystal-clear block of ice in your old-fashioned," she told me.

This trend among bartenders to serve the clearest ice possible isn't all that new. Weatherup, like Petraske before her, is taking part in a centuries-long tradition that considers mixology not just an interesting practice but an art—even a performance. Not the kind of exaggerated, acrobatic performance that Tom Cruise executes in the 1988 movie *Cocktail*, mind you, but one wherein a bartender's costuming, the shape of the glasses used, and the quality of ice cut all play a role in delivering a specific sort of experience. It's a tradition that began where the American cocktail began, in the heart of New Orleans.

W hen Harry Tudor arrived in New Orleans in 1821, ice was still considered a luxury for the wealthy. He changed that by convincing bartenders throughout the city to add frozen chunks of ice to their cocktails, giving patrons their first-ever taste of chilled alcohol.

By the mid-nineteenth century, the family icebox was well established, and demand for ice was great enough that even the poorest parishes of New Orleans were building community icehouses. These large houses resold ice purchased from companies like Tudor's to the citizens of Louisiana, who added ice to spirits drunk at home. In less than fifty years, ice had become critical to imbibing.

In 1868, the city built its first commercial mechanical-ice facility. The steam-powered plant took up more than a block on Delachaise Street and produced up to three hundred blocks of ice per day.

Its thick masonry walls looked from the outside like a fortress. Inside, the walls were lined with cork from floor to ceiling to insulate the blocks and keep them from melting. (The company has since been rebranded as Pelican Ice and claims to be the oldest continuously operating ice company in North America.)

The city's bartenders, unlike their compatriots up north who still relied on polluted natural ice, began experimenting with not only how ice made their cocktails taste but how it made them look. People from all over came to the city to experience the drinks themselves, and the bartenders with the most memorable creations became national celebrities.

The most famous was Henry Charles "Carl" Ramos, inventor of the eponymous gin fizz and a proud teetotaler. Ramos was born in Vincennes, Indiana, in 1856, and moved to Louisiana with his family as a boy. He began his career by learning from others. He started out in a Baton Rouge beer saloon on Exchange Alley, where he learned the basics of mixing booze and ice. In 1887, by then in his early thirties, he and his brother opened a bar in New Orleans called the Imperial Cabinet.

Ramos was tall with a high forehead and sported a neatly trimmed mustache. Nothing about him suggested debauchery, and yet he was widely respected among the city's ne'er-do-wells and churchgoers alike. He was a gentleman of the highest order and closed his bar every evening at the decent hour of eight o'clock to discourage drunkenness. The bar was originally closed on weekends, but after years of pleading from the community, he opened it a mere two hours on Sunday afternoons. Ramos could often be found conversing with patrons to keep an eye out for tipsiness. He cut off anyone who he thought had overdone it. When Carrie Nation, famous Victorian-era prohibitionist and vociferous saloon slayer, visited New Orleans, she did not damage Ramos's establishment. As Elizabeth M. Williams and Chris McMillian write in *Lift Your Spirits*, "She averred that if everyone who served liquor was like Mr. Ramos, there would be no need for the Prohibition movement."

In a city like 1880s New Orleans, where fun and drink often veered into Dionysian madness, Ramos's staid sensibility should have lost him customers. Instead, he became one of the most well-known bartenders in history. Customers clamored for his state-of-the-art cocktails with clear ice cut from blocks made by the nearby ice plant. A retrospective on the bar in a 1928 issue of the *New Orleans Item-Tribune* states that "nobody could get drunk at the Ramos bar, not only because old Henry wouldn't let them, but because drunkenness would take away their appreciation of the drinks."

Ramos's signature drink was the Ramos gin fizz, originally called the New Orleans fizz, which he created in 1888. His recipe called for the sprinkling of powdered sugar, but the secret was in how Ramos used ice. He insisted that the cocktail be shaken with ice for a full twelve minutes before serving, to ensure a frosty pour. On busy nights, the rigorous shaking became too much for any one bartender, so Ramos employed up to twenty men at a time to keep the drinks coming. The men became known as "shaker boys" and worked in rotation to share the burden of making the drinks. In 1900, the *Kansas City Star* crowned the Imperial Cabinet "the most famous gin fizz saloon in the world," adding, "Ramos serves a gin fizz which is not equaled anywhere." By the time Mardi Gras came round in 1915, the drink was so popular that Ramos had to hire fifteen additional men to keep up with orders.

Why, though, did customers clamor for the gin fizz? The spectacle of all those men with shakers was certainly part of the draw, but folks also loved the drink's unique flavors and textures, which developed from all that shaking. As the cocktail expert Dave Arnold reminds us in *Liquid Intelligence,* a cocktail shaken back and forth over ice gets cold fast. The movement also infuses bubbles of air into the liquid, which aerates the alcohol and emphasizes subtle flavors. Both these things—the coolness and the aeration—are enhanced by large pieces of ice, the kind that Ramos was working with in the 1880s.

This was an era before automated ice makers, before the Kold-Draft machines that bars would adopt in the mid-twentieth century to ensure perfectly square ice cubes in every drink. In contrast, Ramos's ice was made in a plant and likely arrived at the bar in 100-pound blocks that had to be broken into smaller pieces by hand with ice picks and hammers. The chunks that went into Ramos's cocktail glasses would have been at least two inches wide, which would have chilled his cocktails quite quickly without diluting them. Simple science tells us why: Ice melts at the surface, so any increase to its surface area leads to an increase in how much and how quickly it melts. Ice also chills at the rate that it melts, so if it melts fast, it will chill quickly, too. Shaking a cocktail brings more liquid into contact with the ice, increasing the surface area of the cocktail and the rate at which it cools. Thus, a shaken drink cools quite quickly.

Larger ice cubes also reduce the amount of dilution in cocktails because they have less surface area overall than several smaller cubes do. Thus, the famed teetotaler would have been serving drinks that were not only colder than those made a century later, but boozier than those at a *Mad Men* lunch.

Ramos served his last gin fizz at midnight on October 27, 1919. He had become an ardent supporter of Prohibition and decided it was time to close the Imperial Cabinet, but not before telling reporters how to make the Ramos gin fizz:

(1) One tablespoonful powdered sugar

Three or four drops of orange flower water

One-half lime (juice)

One-half lemon (juice)

(1) One jigger of Old Tom gin. (Old Gordon may be used but a sweet gin is preferable)

The white of one egg

One-half glass of crushed ice

About (2) tablespoonsful of rich milk or cream

A little seltzer water (about an ounce) to make it pungent

Together well shaken and strained (drink freely)

Ramos further instructed the reporters' readers to "shake and shake and shake until there is not a bubble left but the drink is smooth and snowy white and of the consistency of good rich milk."

The Volstead Act made Prohibition the law of the land on January 16, 1920. Louisiana was one of the first states to ratify the amendment, but the people of New Orleans resisted putting the law into practice for as long as possible. According to Williams and McMillian, "Banning booze was akin to assaulting the city's culture." The city's French, Spanish, and Italian Catholics were particularly unwilling to give up alcohol, because they used it as part of their religious rituals. Others claimed the law had been pushed through by a loud Protestant minority and didn't reflect the beliefs or desires of the majority.

Rather than shutter their doors, bars became speakeasies with hidden entrances and passwords for admission. One of the city's most famous night spots, Tom Anderson's Saloon, was among the first to serve booze illegally after dark, and the Holland House, known for its fine dining, used its restaurant as a shield for the bar in the back. Other establishments opened soda fountains and ice-cream parlors, both of which required an enormous amount of ice to operate.

Prohibition took a heavy toll on bars, but the ice industry was booming. Elsewhere in the country, breweries were getting creative with their ice cellars to survive. According to Maureen Ogle, author of *Ambitious Brew: The Story of American Beer*, more than 1,300 American breweries were in operation in 1915, but only 100 made it through Prohibition. Among the survivors were Coors, Miller, Anheuser-Busch, Pabst, and Yuengling, all of which relied on ice to develop new business models. Some of them shifted to selling malt syrup marketed as a baking ingredient. Anheuser-Busch switched to making infant formula, which could be stored in the company's massive ice cellars. Coors and Miller became leading producers of malted milk, which they kept cool with ice. Yuengling opened a dairy across the street from its brewery and began producing ice cream. (At the time of this writing, Yuengling ice cream is still sold in grocery stores.) And the Milwaukee-based Pabst Brewing Company diversified by selling another popular Wisconsin product—cheese. During the length of Prohibition, Pabst sold more than 8 million pounds of cheese aged in the brewery's ice cellars. After Prohibition was repealed in 1933, Pabst sold its cheese line to a company named Kraft.

A s the cocktail historian David Wondrich reminds us, the early written history of cocktail culture in America is scant because "there was no Homer to record the names and deeds of bartenders." Recipes weren't written down. Historical moments were shaped by anecdotes from patrons passing through and barroom reminiscences. The history of drinking culture is in some ways so ephemeral it might as well have been carved in ice.

In a way, it was.

Wondrich divides the history of American cocktail culture into that which took place before the widespread use of ice, and everything that came after. Before ice, cocktail recipes were simple. Drinks were served lukewarm. Most were rum-based, because rum was easily imported from Jamaica and Cuba. Cocktails were rarely drunk more than a day's carriage ride from the coast, because deep in the New World, key ingredients like sugar, fruit, and mint were hard to come by. In the country's center, if alcohol was drunk at all, it was likely to be rum or whisky—that which could withstand long-distance shipping or be made in a tub—and was served plain in whatever clay mugs were within reach.

But the introduction of ice changed everything. Whether carved, shaved, cracked, or chipped, it became the magic ingredient that transformed cocktails in the United States from after-work indulgences to works of art.

From the perspective of European visitors, the mixing of icy cocktails was the *only* art happening in America. "Who reads an American book? Or goes to an American play?" sneered the English critic Sydney Smith in an 1820 issue of the *Edinburgh Review*. "Who drinks out of American glasses? Or eats from American plates? Or wears American coats or gowns?" The answer was almost no one. America was less than sixty years old in 1820, too new and too rugged to have developed its own aesthetic in any medium that Europeans would have considered respectable. (The drawings and objects made by people who lived on the continent before colonization were rarely considered by nineteenth-century aesthetes to be real art.)

But thanks to ice, an art form in less-than-respectable places like bars was emerging. Behind the bar at the City Hotel in New York City was Orsamus Willard, America's "first celebrity bartender." Born in 1792, Willard was, wrote a patron at the time, "the first in the city to concoct fancy drinks." He was also the first to popularize the use of ice in mint juleps and went on to make the julep his specialty. By the time he retired in 1849, his approach to mixing ice with gin "was perhaps as famous in its day ... as Whitney's automatic cotton carder."

An even bigger star was Jerry Thomas, who elevated cocktail making to a performance, complete with costume. Wrote one reporter who observed Thomas behind the bar:

> He is a gentleman who is all ablaze with diamonds. There is a very large pin, formed of a cluster of diamonds, in the front of his magnificent shirt, he has diamond studs at his wrists, and gorgeous diamond rings on his fingers—diamonds being "properties" essential to the calling of a bar-tender in the United States.

The diamonds (or more likely gemstones) resembled the ice in patrons' glasses. They were sparkling reminders that cocktails made with ice were a luxury and a fine art.

Thomas was born in 1830 in Sackets Harbor, New York, then moved with his family in the 1840s to New Haven, Connecticut, a city of sailors and college boys, and thus a fine place to learn the art of serving booze. He died in New York, with gambling debt and having pioneered another distinctly American art form, the minstrel show. In between, Thomas perfected his cocktail making in San Francisco, a city of barely more than a few hundred tents and streets of mud. Its population was a mix of New England farmers, midwestern bankers, indentured laborers from the Pacific Islands, and southern plantation owners' sons. Prostitutes from France offered their services to men in saloons, who paid with pinches of gold dust. The gold rush was on, and the people of San Francisco were burning with gold fever.

Thomas was one of them. He arrived in San Francisco at the age of nineteen with a pickax and little else. He spent most of his days in the mountains digging for gold, but in the evening, he'd serve drinks from behind bars in saloons consisting of four rickety walls and roofs made of canvas torn from the tents of forty-niners. The clientele was rough—fights in the saloon broke out nightly—but the drinks were exquisite. Mint juleps, gin fizzes, hard lemonades (simply "lemonade" then), sherry cobblers, whiskey sours. Each was served over ice shipped from Alaska. Thomas, the showiest of the city's barkeeps, dipped freshly cut ice cubes in gold dust before dropping them into cocktail glasses, where they'd crack and hiss under pours of whiskey and gin. Even the bar's most frequent patrons turned to watch him make the evening's first icy drink.

Ice turned Jerry Thomas into a magician with the ability to conjure flavors unlike anyone had ever tasted before. The power of ice to dilute altered the notes of familiar spirits, creating opportunities for mixers like lime juice, egg whites, and sugary syrups. As for syrups, they became popular in the making of shaken drinks, where rapid cooling prevented granulated sugar from fully dissolving. (The adoption of syrups also brought an end, thankfully, to something called the sugar drawer, a small pullout behind the bar from which bartenders would spoon sugar crawling with insects.)

News of Thomas's icy cocktails spread east as gold diggers returned home. Then, in 1852, he followed suit. He had made a staggering amount of money ($300,000 in today's dollars) and could have made the journey in about a month by steamboat. Instead, he traveled with a band of Argonauts through Mexico, where they were nearly shot by a mob. When he finally made it to New York, he opened a bar with George Earle under Barnum's Museum, one of the most popular attractions in the city. But for reasons lost to history, he left for a job in New Haven, then returned to New York to run several other bars. He and his brother eventually opened a saloon of their own, and for a while, it was one of the most famous in America. Thomas published two books on cocktails and got married. Everything pointed to a happy life, except for one thing: he had developed a significant gambling habit and lost most of what he made. By 1876, Thomas was broke. He died nine years later of heart disease.

Thomas's story might have ended there if not for the rediscovery of his two books nearly a hundred years later. Those books taught modern bartenders how to make cocktails without commercial mixers and, crucially, with various sizes of ice. Today when bartenders experiment behind the bar, they're thinking as much about ice as mixers and spirits. In essence, Thomas helped usher in a new ice age.

From 1955 until sometime in the late 1980s, Kold-Draft machines could be found in every bar, hospital, and restaurant in America. The machine made perfectly shaped, one-and-a-quarter-inch cubes that came out so cold they would take forever to melt. But the machines broke often. Some bars switched to the Japanese-manufactured Hoshizaki machines, which were known for their reliability, but their little half-moon ice pieces had a greater surface-area-to-volume ratio and melted quickly.

The machines competed for business until the early 2000s, when bartenders rediscovered techniques for cutting their own ice. In an interview, the bartender Chris McLeod of Dutch Kills bar in Queens, New York (co-owned by Richard Boccato), said, "We want all our ice to be beautiful." To ensure that's the case, his team uses a Clinebell machine, the same kind used at Weather Up, to produce 300-pound blocks, which are cut into smaller shapes using a chainsaw. So important is this process of making and cutting ice that Dutch Kills hired a full-time ice specialist. The bar now has an offshoot business, Hundredweight Ice, which supplies packs of cut cubes to over seventy clients throughout New York City.

Bars with similar interests in ice have opened all over the country, giving rise to the kind of ice company that specializes in ice for bars. The Ice Doctor in Gainesville, Florida, for example, offers customized ice to the bars it serves; each cube is cut to the size of each client's specific glassware. Minnesota Pure & Clear sells ice to bars and leases its freezers to bars who need more storage space. Névé Ice in Los Angeles has been selling specially made ice to bars since at least the mid-2000s and has had its cubes featured in episodes of *Mad Men*. Miami's Mixology Ice has cornered its home cocktail market and is looking to sell its ice nationally. It's already selling ice to bars in New York and Washington, D.C.

All this fancy ice is expensive to make and purchase. It's also hard on the environment. A Clinebell has to run for at least three days to make enough ice to serve one night's worth of customers at a typical New York City bar. None of these machines have energy efficiency ratings. Artisan delivery companies, meanwhile, are bringing their product to customers in vans, which spew carbon into the atmosphere.

There is an alternative to all this, which is making clear ice the old-fashioned way—by using the "directional freezing" method at home. The term was coined by ice expert Camper English, who has been fascinated by ice since he was a boy. "When I was a kid," he told me, "I was told to boil water because it makes the clearest ice. Everyone was told that. I guess no one tried it, because it doesn't actually work." English would know, because he's tried everything to make clear ice, he tells me. Boiling it, freezing and refreezing it. He even tried freezing carbonated water to see if the bubbles affected the ice's clarity. They didn't.

"Eventually I figured out something pretty obvious," he said, "which is that ice freezes from the top down." As Frederic Tudor knew well, as ice freezes, it pushes all the water's impurities to the bottom. The top part is left as clear as glass. With the science of natural ice in mind, English filled an insulated container without a lid and put it in the freezer. The water froze from top to bottom, pushing the impurities into the lower one-third of the block. A day later, he took out the container, popped out the ice, and, sure enough, found himself looking at a small block of ice whose top two-thirds were exceptionally clear. He cut off the cloudy bit and sliced the clear part into sparkling cubes.

He insists the ice he makes in this way isn't hard to cut. "People shouldn't be intimidated by cutting ice," he told me. "It's the fun part! You can do it using any number of tools around the house. I initially used a wood saw. That's really inefficient and unnecessary. Now I have a whole series of ice picks. But you can use a bread knife, a mallet, whatever you have. It just depends on how important straight lines are to you."

The directional freezing method has caught on in bars and restaurants all over the country, from New York City, where space is at a premium, to smaller markets where Clinebells are hard to come by. English has also inspired a generation of ice enthusiasts who use the method to make clear ice at home. "Ice Instagram is pretty wild," he said. "It's just everywhere now."

When I asked him if he had a favorite kind of ice, he said no. He finds it all fascinating. But when he walks into a bar, he can tell how the ice in his glass was made just by looking at it. "Yeah," he said with a self-deprecating smile. "I'm great fun."

CHAPTER 8

Fire and Ice

The St. Paul Winter Carnival Ice Palace (1887)

Before there were amusement parks, before there was Netflix, people in dire need of entertainment went to dinner.

In medieval Europe, villagers gathered at large banquet halls to celebrate holidays or toast victorious soldiers. Around A.D. 1100, one of the world's first restaurants opened in China. For centuries, in almost every society on the planet, drinking and eating have been considered the first steps toward merriment, so it's no wonder that some modern-day chefs strive to cook and entertain in equal measure. They plate their culinary concoctions to look like sculptures. Some serve their meals with little cards that explain what inspired each dish. And many chefs have learned to sculpt ice.

With a startling level of precision, chef-sculptors wield chainsaws and ice picks to transform 300-pound blocks of crystal-clear ice into swans, horses, teddy bears, and baskets of fruit, which become centerpieces for wedding buffets, graduation dinners, milestone birthday parties, and high school reunions. Once, at a New York City gala for a prominent arts institution, I watched a chef put the finishing touches on a four-foot-high ice carving of a whiskey bottle. It held its shape for hours, even as the air inside the ballroom warmed as hundreds of guests took to the dance floor.

Ask any ice sculptor working in America today where and why they learned to carve ice, and you'll hear a similar story. The sculptor went to culinary school, found a job in the restaurant industry (or perhaps still works in the industry), and met another chef who knew how to sculpt ice. The experienced sculptor took the less experienced chef under their wing and, over the course of several years, taught them how to transform blocks of ice into art. It's an apprentice model used by everyone from wood artists to electricians to plumbers. Except, in the case of ice sculptors, the path toward learning to carve is littered with dozens of cracked blocks of ice.

The practice of sculpting ice didn't begin with American chefs. The Inuit have long built homes packed with snow and carefully carved ice. Eighteenth-century Russians sculpted ice as part of a New Year's Day ritual, wherein ice harvested from lakes was carved into crosses believed to scare away evil spirits. Those unlucky spirits then got sucked through the holes where the ice used to be, back to the underworld where they belonged.

Ice sculpting as it exists today, as a form of spectacle (and even competition), came to the United States in the early 1800s, a time when ice was still a symbol of prosperity. Carvings were incorporated into expensive, elaborate dinners, often as part of the presentation of ices and ice creams. The desserts were served in small, glittering bowls carved from ice. Sometimes sculptors would freeze pieces of fruit or spun sugar into their designs to create patterns of color.

Given the connection between ice sculpting and wealth, it comes as little surprise that American interest in the art peaked in the mid-twentieth century, just after World War II, when the country was experiencing an unprecedented economic boom. For the first time in decades, Americans had money to burn. And now that the war was over, it felt appropriate to celebrate again. So they did—with ice. Ice sculptures were everywhere, in restaurants, at large parties, even in movies. In *Citizen Kane*, an ice carving at the center of an upper-class party comes to symbolize the host's cold treatment of his employees.

In 1947, August Forster published a foundational work on ice sculpting, *Fancy Ice Carving*. Its lessons include how to shape ice into gondolas the length of a dining room table, cornucopias that hold dozens of appetizers, and a mini sailboat for serving fish.

Ice sculptures continue to fascinate. They still feature prominently at galas and other elaborate celebrations. Ice companies across the United States make blocks of ice exclusively for ice sculptors. And ice sculptures have continued to appear in movies, though their connotations have evolved. Instead of symbolizing coldness, they now suggest romance. In *Edward Scissorhands,* the titular character carves with his scissored hands an ice statue for the woman he loves. In the 2015 Hallmark movie *Ice Sculpture Christmas*, a novice chef falls in love after entering an ice-sculpting competition.

Why has the popularity of ice sculptures endured? They're beautiful, sure, but they don't last. Perhaps that's the draw. Their ephemeral nature demands focused appreciation, which in turn imbues them with specialness. Enjoy them now, because in just a few hours, they'll disappear forever.

I wanted to learn more about how these fleeting works of art get made, so I went to Cheshire, Connecticut, to visit Bill Covitz, ice sculptor and founder of Ice Matters, a company that specializes in ice sculptures for events. Covitz's sculptures have adorned tables at children's birthday parties, weddings, and boozy gatherings of old friends for over a decade. He's also an ice-carving champion who's competed against sculptors from around the world. Covitz has even worked as part of a team that carved musical instruments out of ice for Norway's annual Ice Music Festival. Every year, musicians gather there to play cellos, guitars, and marimbas made from ice. (The festival used to be held in Geilo but, in recent years, was moved to Finse because Geilo became too warm.)

I pulled up to Covitz's white barn on a bright November afternoon, three weeks before Thanksgiving. The day was unseasonably warm, but Covitz greeted me outside wearing ski pants, a wool hat, and thick gloves. After some pleasant hellos, he took me inside the barn and to the door of a small room his team calls "the office." The room was one of at least two insulated ice lockers on the premises that he keeps at a toasty fifteen degrees Fahrenheit year-round. When he opened the heavy door, a blast of cold air hit my face, and I was immediately glad that I'd taken to heart his warning to "dress warm." I pulled my scarf tighter around my neck and followed him in.

Inside, I could see my breath. I could also see a lot of ice. Along the walls were blocks of ice approximately forty inches tall, twenty inches wide, and ten inches thick, each made on-site from Clinebell machines that run several hours per day right there in the barn. I noticed a frozen Butterball turkey still in its wrap resting on one of the blocks. When I asked Covitz about it, he shrugged. "Got ours early this year."

In the back of the room was a machine about four feet high and two feet wide. Inside was a small ice block standing on end. "This is a CNC machine," he said. His tone was both proud and dismissive, for reasons I was just about to understand. A CNC (or computer numerical control) machine is a relatively new technology in ice sculpting. It utilizes a laser to cut 2D designs into blocks. The machines are most often used for cutting letters and commercial logos, because the laser can cut straighter lines than even the most experienced sculptor.

"Problem is," Covitz said, "some people *only* use these." The CNC, he continued, has put some sculptors out of business. Customers couldn't see paying for a hand-carved, 3D sculpture when they could get a 2D one made by a CNC for much less. He finally purchased a CNC because that was the only way he could compete for business. It was a bold move that almost cost him a friendship, when the friend, a fellow sculptor, briefly stopped speaking with him. "He finally came around," Covitz told me.

Covitz's business has survived largely from commissions from municipalities and universities that sponsor holiday events. And one of those events, the annual Holiday Stroll in Ridgefield, Connecticut, was fast approaching. He had been hired to carve several ice sculptures that would be displayed up and down Main Street as part of a holiday-themed scavenger hunt.

"Want to see me carve the reindeer?" he asked.

"Absolutely."

I followed him into a second ice locker, this one larger than the first. A loud fan was blowing inside to circulate the air and keep it cold. Along the wall stood several sculptures already finished. He picked up a frozen teddy bear, maybe two feet high, and held it out to me so I could get a better look. "This one's for a birthday party," he shouted over the fan. He set it down and gestured toward an elaborate sculpture featuring a small slide that snaked from the sculpture's top to its base, where it widened like a river delta. "This is a shot flue," he told me with a laugh. "A bunch of guys get together for a reunion every year and ask me to make this." When he tried to pull what looked like a horse away from the wall, he accidentally dinged it, sending a small piece of ice flying. "Damn," he said. "I'll fix it later. That shit happens all the time."

Across the room was a workbench, where he'd lined up his carving tools like a surgeon's. Most carvers utilize six or seven differently sized chisels for chipping and gouging the ice. Covitz had at least this many as well as numerous chainsaws and a die grinder for detailing. Ice tongs to lift the blocks dangled from hooks in the ceiling. The room's resemblance to a torture chamber was not lost on me.

He picked up his chainsaw. "Ready to see some snow fly?"

I nodded, and he waved his hand toward the back wall, the universal sign for "step back." I moved away as he walked toward what looked like a pedestal. On top was an ice block standing on end. His chainsaw roared to life. Five minutes later, I was shaking from the cold, but I didn't ask to leave. I could hardly believe what I was seeing. In that short period of time, and with just a few precise strokes, Covitz had carved an outline of a reindeer.

Next, he picked up a chisel and refined the shape. Then using a die cutter, he added lines and dots that quickly became fur and a face. In a mere twenty minutes, the body of the reindeer was finished. He told me that he'd carve antlers out of a separate block on a later date and then use a small torch to weld the antlers to the reindeer's head. Before we left the room, he stuck a red, battery-powered light on the reindeer's nose. So this was Rudolph.

The sculpture was beautiful, but it also looked quite fragile. I asked him how he'd get it to downtown Ridgefield, a forty-seven-mile drive away. He told me he lined a refrigerator truck with thick blankets. "We set the carvings up an hour before [the event]," he said, "because they frost up." He or someone on his team would have to take that same small torch and melt all his sculptures' outside layers just before showtime to make them shine again. The mental image of flame on ice made me think of these lines from a Robert Frost poem:

> Some say the world will end in fire,
> Some say in ice.

But here, fire and ice weren't taking turns destroying the world; they were working together to make it more beautiful.

Outside the barn, we sat at a wooden table in the sun. Covitz peeled off his hat and coat, but my teeth were still chattering, so I kept mine on. I asked him how he became interested in ice sculpting, and he told me his career, like those of so many other ice sculptors, began in the restaurant business. He'd met a chef-sculptor who took him under his wing, and Covitz, it turns out, was a natural. "I don't draw designs ahead of time," he told me. "I just look at the ice and see the sculpture inside it."

Before I left, I asked him for his thoughts on the origins of ice sculpting. He said that no one knows for sure where the practice began, that it probably has several origins, including those from Indigenous communities. But one interesting inflection point in the history of ice sculpting, he said, was the rise of ice palaces in the nineteenth century. These were full-scale, multistory castles carved from ice harvested from lakes and streams. People would visit them from hundreds of miles around, and on the last night of the castles' existence, just before the spring months arrived to melt them away, a staged battle would take place, complete with fireworks and faux cannon fire. "They used to really blow those places up," Covitz said.

Fire and ice, indeed.

———————

"Looking through the eyes of love" goes the theme song from the movie *Ice Castles*. That movie was actually about ice *skating*, but the song is more historically relevant than the songwriters probably knew. The first ice castles had everything to do with love. They appeared in Russia in the eighteenth century as honeymoon suites. The newlyweds would arrive at the castle on elephant-back, followed by a procession of sleds pulled by reindeer and hogs.

The castles made their appearance in North America a century later. The first went up in Montreal in 1883 as part of that year's winter carnival, a celebration of ice and snow that helped pass the longest and darkest nights of the year. The city hired Alexander Cowper Hutchison, one of Canada's most prestigious architects, to design the castle. The hope was that the structure would draw large, international crowds and convince them that Montreal was no longer the small, quaint town it once was. This was, after all, the year that the Brooklyn Bridge opened in New York City. Contemporary engineering was producing marvels, and Montreal wanted to compete for the prestige and tourist money that such marvels wrought.

The city had some work to do in that regard. An American reporter had recently described Montreal as having a "quaint old market-place, where habitants in odd-looking costumes sell their meat and vegetables, and fat old French Canadian women barter frozen cream and eggs, a strange preparation of corn boiled in lye, and many other delicacies in name and manufacture incomprehensible to the stranger." The ice castle, thought the city's officials, would convince even those snotty Americans that Montreal was a modern city.

It did and it didn't. In its final form, the 1883 castle was a 90-foot cube made of 200-pound blocks of ice stacked on top of one another, with a 50-foot tower rising at each of the four corners. A center tower rose to 90 feet. All the towers boasted pointed roofs made of wood. The wood was sprayed with water, which froze into sheets of icicles. Sixteen electric lights illuminated the castle at night. Electric light was still a novelty in those days, and a reporter who visited the castle wrote that the lights "were enough to make it gleam and glint in the darkness of a winter's night, like a fantasy out of the Arabian night."

The castle was built a century before the *Super Mario Bros.* video game was created, but photos of the castle look to modern-day eyes like that pixelated, early-'80s Nintendo castle made of digital squares.

Its "windows" were tiny slits in the ice. Its "archway" entrances were more rectangle than parabola. The castle definitely drew a crowd, but not to the extent that the Montreal government had hoped. The city tried building an ice castle every year until 1886, when a smallpox epidemic struck, killing more than two thousand people in Montreal in mere weeks and putting the winter carnival on hiatus.

That was also the year that the tuxedo was introduced in Tuxedo Park, New York; the Statue of Liberty was unveiled; and St. Paul, Minnesota, became America's fast-growing city, fueled by railroads that brought immigrants and established the state's agricultural economy. Its population had swelled from 39,000 in 1880 to 120,000 just six years later. Yet, like Montreal, St. Paul was suffering an image problem. A New York newspaper journalist had described the city as "another Siberia, unfit for human habitation" in winter. Rather than prove the journalist wrong, St. Paul embraced the description. It established the St. Paul Winter Carnival Association and chose the city's Central Park for the site of its own ice palace, in the tradition of Montreal's successful build. It even hired the same architect, Hutchison, to do the designs.

Within the carnival grounds was an encampment of seventy-five members of the Sioux Nation, who performed their "Indianness" (another marvel for white visitors) by simply sitting and existing. The carnival also featured an exhibition hall and rinks for curling and ice skating, two sports that had come to prominence in the United States only in the previous two decades, thanks to mechanically made ice.

The St. Paul castle stood fourteen stories high and was constructed out of more than thirty thousand blocks of ice cut from the Mississippi River and nearby Lake Como. Twenty ice-hauling crews using horses and large sleds pulled the blocks from the river and lake to the construction site. When ice ran low, other towns, some as far as the Dakotas, donated their own ice. In its final form, the castle measured 106 feet high and 180 feet wide. The outer walls measured 20 inches thick. Four grand arches served as entrances to the palace's interior. Inside the castle was an imposing hallway with a 40-foot-high ceiling. The hall led to several rooms, each 15 feet square and containing enormous ice statues. Some of the rooms connected to stairways that led to the ramparts above. Like the castle in Montreal, this one was illuminated at night with colorful electric lights.

The St. Paul Winter Carnival drew hundreds of thousands of visitors from across the country and Canada. It was the first to establish one of the stranger ice-castle traditions: the "storming of the castle." On its last night, the ice castle was "stormed" by actors performing an imagined battle. The "generals" leading the "soldiers" explained to audiences that the battle was justified because they had to protect it from an angry god. King Borealis of the North and his queen, you see, had left their home in "the icy caverns of the crystallized region beyond the ken of man" to come live in this ice palace. The Fire King had apparently grown jealous of their new digs and began attacking the castle. The reenactment included fireworks exploding over the glistening palace. That first year, the storming alone attracted fifty thousand spectators.

The following year, St. Paul hosted another winter carnival with an even larger palace, this one designed by Charles E. Joy. It towered over the city at 217 feet high and featured flying buttresses that reared up to several small flanking turrets. Visitors entered the ice palace through an archway 16 feet wide and 15 feet high, and inside was a skating rink surrounded by several ice statues. Above the archway sat a colossal statue of King Borealis holding a bright electric light. His sides were flanked by carvings of polar bears. Nearly two hundred electric lights lit the palace at night.

Every year, the St. Paul carnival grew larger, and so did its central ice palace. Inside the 1888 palace, a wedding was performed in front of six thousand guests. The couple arrived escorted by the local police, a marching band, a drum corps, and the local snowshoe club. When they took their vows, a red flare shot through the sky. The palace's interior was adorned with statues of a polar bear, a skater, a curler, and a tobogganer. It also housed a maze of about 80 feet in diameter. The association planned to hold three stormings that year, and the first two went off without a hitch. But on the night of the third, the temperature rose to the forties and the castle began to melt. The fireworks that had been stored inside were soaked and left to rot. There would be no angry gods at war that night.

Warm temperatures continued to thwart the St. Paul carnivals, and the Winter Carnival Association put them on hold until 1896, and then again until 1916. Now the twentieth century, the event sought to distinguish itself by utilizing a newfangled gasoline engine in its nightly procession. Many of the motorized floats broke down, however, and the association banned them from the following year's parade. In 1917, twenty-one thousand people, half the size of the entire U.S. Army at that time, enrolled in a marching parade. That spectacle was shadowed, however, by a 510-mile dogsled race, whose winner was deemed a hero by the American press after his lead dog died and he completed the race on foot. The carnivals in St. Paul ended in 1918, when the United States entered World War I, and the nation's battles were no longer staged.

Other cities attempted to construct ice castles, and most of these attempts failed miserably. One such town to give it a go was Leadville, Colorado. At an altitude of 10,200 feet, Leadville is the nation's highest incorporated city, but in 1896, it was little more than a rough-and-tumble mining town frustrated at having to live in the shadows of glitzy Denver. The Panic of 1893 had brought an end to the town's fledging mine-based economy, and the repeal of the Sherman Silver Purchase Act caused the prices of ore—its main export—to plummet. The men of Leadville were out of work, and the women who lived in the once prosperous whorehouse found themselves without any clients. Up and down the town's few dusty streets, Leadvillians were packing up and heading out.

For the saloonkeepers and gamblers left behind, the St. Paul ice palace seemed like a golden ticket to revitalization. They banded together to form the Crystal Carnival Association and raised more than $40,000 from the sale of association stock. This carnival, thought the top brass of Leadville society, would show the state of Colorado just how refined their tiny town could be. "People in the valleys," read a local newspaper of the day, "are beginning to realize that Leadville is something more than a rude mining camp. The cloud city has made her entrée, at last, into society circles." Their highfalutin hopes were dashed, however, when a few days after the article appeared in print, a local named Lou Bishop was arrested for riding his horse through a saloon.

Still, the Crystal Carnival Association remained undaunted in pursuing its dream of a Leadville ice cathedral. It hired St. Paul's architect C. E. Joy to build the palace, but one with parameters that made this the most difficult one the architect would ever build. It demanded that this be the largest ice palace ever built, that a permanent wooden pavilion inside the ice structure be created to be reused in subsequent years, and that all of it be ready by Christmas. Most winter carnivals opened in late January or early February to allow their builders more cold-weather time to complete construction.

Approximately 350 men worked on the Leadville palace, while local contractors supervised the interior woodwork. Additional men sawed ice blocks from the Arkansas River and Palmer Lake and hauled them by horse and sled to the construction site. The palace used more than 5,000 tons of ice.

The men were thwarted at every turn. The weather that December was unusually warm. On December 12, the temperature in Leadville was as high as sixty-five degrees Fahrenheit. The ice was melting faster than the men could lay it, and a strong wind descending from the eastern slopes of the Rocky Mountains cut crevices through the walls. The men draped swaths of canvas over the structure to shade it during the day, and at night, the fire department sprayed the entire palace with water to fill in the cracks and seal the melting blocks together.

Bad weather was just the start of the builders' problems. The lift that was used to move the ice blocks broke and fell, severely injuring one of the men. Not long thereafter, one of the towers collapsed, bringing a wall down with it. A few days later, two men fell from another tower, tumbling several feet to the ground; contemporaneous accounts suggest they lived.

The palace wasn't finished in time for Christmas, but it opened on New Year's Day in 1896, with a costume parade in the afternoon and skating in the evening. More than 2,500 people visited on opening day. The carnival brochure suggests that the castle was 435 feet long and 325 feet wide, with the tallest towers reaching 90 feet, but photographs suggest they were no higher than 60 feet. The size of the ice blocks used to construct the castle are described in some places as 5 by 2 feet and elsewhere as 20 inches by 30 inches. The thickness of the palace walls is described sometimes as 5 feet and elsewhere as 8 feet. It's not just the size of the palace that remains a mystery. No one knows how much it cost, either.

"Nothing ever appeared in print that could be considered a

complete accounting of the Leadville Crystal Carnival Association,"
reads a 1977 newspaper account of the carnival, "nor have the corporation's records ever been found." Estimates ranged from $35,000 to
$140,000, but even that lower figure is ten times the cost of any pre-
vious ice palace in North America.

Some of that money, at least, went to constructing an impressive
entrance, which was adorned with an ice statue of a woman standing
on a pedestal and pointing east. According to the brochure, her fingers
pointed "to the rich mineral hills from whence Leadville's wealth is
taken." The statue may have been intended to serve as an ode to that
which put Leadville on the map, but in retrospect, one can't help wondering if she was pointing toward the exit.

The palace did not draw the crowds the association had wished
for, and those who did come spent little. By February, the weather was
warming, and by March, the weather was so warm that the palace
began to melt again. The storming of the palace, complete with fireworks, was canceled because of an actual storm. A few days later, the
palace closed for good when the town's last remaining miners went on
strike. No other ice palace was ever planned in Leadville.

———————

In 1937, America was once again in an economic depression, this one
worse than any that had come before. Across the Atlantic, Hitler
was conspiring to take over Austria, and Americans spent their evenings by the radio, listening to President Franklin D. Roosevelt's
"fireside chats" in anticipation of going to war. This year, St. Paul decided to revitalize its winter carnival. Supported by the federal government's Works Progress Administration, the city pulled together
the money to construct an art deco ice castle illuminated with
spinning electric lights.

Built opposite the state capitol, the ice castle was smaller than its predecessors, at only 193 feet high, but its construction provided a much-needed distraction for Americans. While work crews put in sixteen-hour days to finish the building, crowds of nearly one hundred thousand people gathered around the site, having come from several states away. The ice castle was such a success that the city built one every year until 1947. The outbreak of World War II demanded ice plant workers utilize their skills and materials for war, so natural-ice harvesting regained popularity, providing not just more ice for domestic use but also jobs. Extra men were hired to cut blocks from nearby lakes to build the ice palaces.

By the late 1940s, postwar America no longer felt a need for ice palaces. The spectacle they once provided was superseded by the excitement of returning soldiers, who were eager to start families. Still, the ice castle left us a legacy of a rich archive of beautiful photographs—and inspired one of the stranger footnotes in military history.

In early 1942, while St. Paul was considering whether to hold another winter carnival, an Englishman named Geoffrey Pyke was finalizing a proposal to the Allied forces. His proposal was to construct giant aircraft carriers out of ice. The plan was appealing to engineers and military officers alike, because carriers made of ice could be built larger than ones made of steel, allowing for the use of faster and better-armed planes. Moreover, ice was much cheaper than metal and wood, which were in short supply. And best yet, an airbase made of ice couldn't be sunk. President Roosevelt thought the proposal a waste of time, but Winston Churchill demanded that the research be made a priority. For the first time in modern history, ice was systematically and intensively studied by military forces.

Ultimately, the ice carrier was never built. Through their research, engineers discovered that ice was unreliable as a structural material—the sun could melt it, and while it couldn't technically be sunk, bombs could fracture it worse than concrete. In early 1944, the project was abandoned.

A little over a decade later, the U.S. Army decided to build a large nuclear-powered camp on the ice cap in northern Greenland. Camp Century spread over an area of 1,400 by 1,000 feet. Prefabricated buildings were erected roughly 25 feet below surface level, because annual snowfall for this area was at least 4 feet. The scientists there noted that snow cleared from roads by rotary plows gained significantly in density, hardness, and bearing strength as a result of having been handled by these plows. The army began to consider whether compacted snow and ice could be used as building materials. They built domes of ice up to 36 feet in diameter to test their theories and a parking lot that could hold up to eleven thousand cars.

The U.S. forces were some of the most well funded in the world, but through these experiments, ice came to represent more than wealth, as it had in centuries past. It had become a symbol of fiery military might.

PART 3

ICE SPORTS

CHAPTER 9

Fancy Figures

**Norwegian skating champion Sonja Henie training
at the Palais des Sports in Paris (1936)**

Long before Frederic Tudor convinced Americans to drop chunks of ice into their cocktails, ice was, in many parts of the world, something less to celebrate and more to overcome. In what is now known as North America, Iroquois Indians once taught French trappers to cross frozen rivers by tying bones to their feet—one of the earliest known versions of the ice skate. Ancient Scandinavians worshiped Ull, the god of hunting, who wore "skates" made of magic bone, they believed, that allowed him to cross the widest of seas. In twelfth-century England, young men striving for knighthood engaged in combative skating—essentially, dueling on ice—to catch the eyes of royalty. Four hundred years later, during the Dutch revolt against Spain, Spanish soldiers were forced to retreat from their camp on a frozen lake when Dutch musketeers on skates took them by surprise. The Spanish soldiers' spiked clogs were no match for the fleetness of the Dutch attackers' bladed shoes. This is the only known military battle to take place on ice skates.

No one knows for sure when *figure* skating—the style we know today as a competitive sport—developed in North America, but it likely came from Europe, where skating had less to do with dancing, which is what we associate it with today, and more to do with the carving of actual figures in the ice. European skaters prided themselves on maintaining rigid postures, as if their upper bodies were frozen solid. Instead of skating to music, they skated to callers who'd shout out shapes for the skaters to create. When figure skating arrived in Canada sometime during the late seventeenth or early eighteenth century, skaters seeking to experiment gave the sport some fluidity by developing "grapevines," elaborate shapes created by loosely crisscrossing the legs.

As skating spread southward into the United States, even more inventive skaters embellished the pastime further with bent legs, turns, loops, and spins. Critics deemed the new American style "fancy skating" to distinguish it from the rigid style still practiced in Europe, where stiffness equaled elegance. To keep the body as still as possible was to be "poetry in motion," wrote Lord Henry Cockburn, member of London's first skating club. In one less-than-flattering comparison, he likened skaters moving opposite each other to a "well-matched pair of horses in harness" that "keep perfect time, step well together, and work harmoniously in concert."

Eventually, using ice-making technology based on John Gorrie's original invention, engineers would bring ice skating indoors, further popularizing the sport (along with hockey, speed skating, and curling) and giving rise to America's first multi-entertainment complexes, like Madison Square Garden. But even before mechanically created ice could transform ice sports as we know them, ice skating was gripping Americans like a fever.

Something about the ice—perhaps the free-fall feeling it gave those who glided across it—made eighteenth- and nineteenth-century Americans rhapsodic. By the 1760s, they were skating along the East Coast whenever the winters fell cold enough to freeze the lakes. In Philadelphia, the largest city in America, with nearly fifteen thousand people, the pastime drew thousands of participants at every rink. Thousands more could be seen making figures on the frozen Schuylkill, wherever the ice was thick enough to support their weight. Ice harvesters had to protect their fields, raising ice saws in the air to dissuade encroaching skaters.

The ponds in New York likewise drew thousands of skaters. The Central Park skating pond alone attracted seven thousand people at any one time. An article in the *Daily News* quipped, "If it had been possible to set the ice up edgewise and skate on both sides of it, every square yard would have its steel-shod tenant."

As one of the most popular places to gather and socialize in winter, skating ponds became petri dishes for shifting social mores, especially those related to gender. At first, women were unwelcome at skating rinks, but that soon changed. By 1867, the American writer Henry Chadwick noted a shift in attitudes toward not only gender but also social class:

> Ten years ago a lady on skates was not only a rare and novel sight in this vicinity, but any fair one . . . who in such a way would have dared to brave the opinion of "her set" and to have outraged their sense of feminine propriety by appearing in public on skates, would have been driven forth from the sacred circles of the then fashionable coteries of the city in disgrace. Now the very reverse is the case.

This was a startling departure from what was happening in Europe and Great Britain, where women—especially upper-class women—rarely took to the ice. On a visit to New York, the English writer Charles Wolley fantasized on his way to Central Park about the number of women he was about to see skating there: "What a charming thing it will be to see five hundred cherry-cheeked healthy beauties—goddesses in crinoline and mortals in plumptiduinous loveliness—gliding, whirling, and now and then sitting down, without exactly intending it, on the slippery ice."

If Wolley seemed a bit randy at the thought of women skating, he certainly wasn't alone. Sex was one reason why skating became so popular in America. The pastime allowed for more relaxed interactions between couples. Men and women could flirt and touch in ways that weren't permitted in other contexts. As one reporter for the *New York Times* wrote, "The girls skate with one gentleman all the afternoon, beyond the sight of mama, who would be nearly frantic were the same thing to occur in a ballroom."

Skating also permitted nineteenth-century women to hike up their floor-length dresses. Those heavy petticoats restricted movement on the ice, so women skaters eschewed the traditional garb for shorter dresses that freed their legs and exposed their feet and ankles. Writing for the *Times*, William Howard Russell spent a day watching women at a skating rink, admiring "legs of every description, which were generally revealed to mortal gaze in proportion to their goodness." He even feigned shock at women dusting ice flakes from their clothes, which included a "dandy jacket and neat little breeches—yes, they wear breeches, a good many of them." A social etiquette book of the day recommended women skaters choose "rich, warm materials" on the ice, and acknowledged that fur trimmings could be used to draw attention to the body parts they adorned.

Thanks to ice, hiked skirts became the fashion everywhere. Women employed a series of rings called "elevators" under their skirts to hoist the petticoat above the ankles. This contributed to the nineteenth-century hypersexualization of feet, a phenomenon heard and seen time and again in popular songs, poetry, and cartoons of the day. Satirizing the trend was the popular ice-skating ditty "Matilda Toots, or You Should Have Seen Her Boots," published by H. De Marsan, about a woman who goes crashing through the ice, landing upside down so that only her feet are visible: "The water next came bubbling up! Crash! I saw the boots / alone, above the water where had gone down Tilda Toots." The drowning woman is rescued by a man who proposes marriage on the spot. She wears her ice skates on their wedding day.

The sheet music sold thousands of copies, its popularity due as much to its cheeky cover as to the music itself. It depicts a woman languishing head-down in broken ice, her skate-clad feet the only part of her that's visible, while men of all social ranks scramble over one another to get a better look at them.

Ice skating did more than help people find love. It democratized the classes by making everyone, no matter their gender or economic status, slip and fall. The 1868 edition of the Brooklyn Skating Rink Association's handbook bragged about this fact: "Skating, in a moral and social point, is particularly suited to our republican ideas as a people. The millionaire and mechanic, the lady of fashion and those of humbler rank, all meet together to enjoy this fascinating and beautiful exercise." Another handbook of the period puts a finer point on it: "No parlor etiquette can be maintained on the ice-pond."

Skating clubs mushroomed up across the upper Midwest and along the East Coast as a means of formalizing the hobby. One in Boston became a charter member of the U.S. Figure Skating Association, which led the way to bringing figure skating to the Olympics. Another in New York, located at McMillan's Pond, doubled its initial membership of 150 in just two years.

Out of these clubs emerged a bold, mustachioed skater named

Jackson Haines, whose audacious choices on the ice led many historians to call him the father of modern figure skating. Born in New York in 1840, Haines trained as a ballet dancer and studied theater as a boy. During the 1860s, he took his dancing to the ice, where he concentrated more on body position than on the patterns he skated. He found the callers dull and invited musician friends to the rink to play while he danced around the pond. His fellow skaters found the sight of a grown man skating to music so ridiculous they routinely laughed him off the ice. This was, critics thought, fancy skating at its worst—showy and imprecise. Haines thought otherwise.

He packed up and left for Europe, landing in Vienna at the height of what's now called the Romantic period. This was an era of virtuosos and star performers with a penchant for big emotions and intricate art, the home of the violinist Niccolò Paganini and the pianist Franz Liszt. There Haines found kindred spirits in the country's many dancers, poets, and musicians. He was especially taken with Vienna's obsession with waltzing and incorporated the dance into his skating. He debuted his waltz on ice at a performance in 1868 attended by Emperor Franz Joseph I, where one American spectator remarked that "he shot in on a long outside [spiral] which took in the whole circumference of the area, performed a pirouette and took off his hat to a Grand Duke."

The performance caused a "great sensation," and the Jackson Haines waltz became the most performed ice dance of the century. Haines toured Europe, becoming skating's first matinee idol and creating more elaborate routines. He devised the sit spin—one of today's most easily recognized skating moves—and would on occasion add stilts to his skates. Sometimes he wore cartoonish costumes. One night, at a show in Vienna, he asked a pupil named Franz Bellazzi to join him on the ice. The two skated in perfect unison, inspiring standing ovations that went on for twenty minutes. The pair kept at it, developing more complex routines, including one wherein Haines took to the ice dressed as a bear and Bellazzi as the bear's trainer. These performances were among the first recorded instances of paired same-sex skating, leading some historians to call Haines an LGBTQ pioneer.

So many people sought Haines's instruction that he set up a school in Vienna. The style became known as the Viennese style and then the International style, which is considered the foundation of competitive figure skating today.

Much of Haines's life details are lost to history, and some of what remains is the stuff of legend and myth. We do know that Haines died young, at the age of thirty-five, possibly thirty-six, and probably of pneumonia (some say tuberculosis). He died in Finland, where the Rotary Club of Gamla-Karleby provided a plaque for his gravesite that reads "In remembrance of the American skating king." A century later, the World Figure Skating Hall of Fame was established and made him one of its first inductees.

Haines never returned to the United States, but his pupils did, bringing the Viennese style with them. "Ice dancing" became all the rage. Musicians played next to skating ponds. Violinists sawed their instruments in the cold, the tips of their gloves snipped off so they could better feel the strings. Ice-skating clubs around the world began holding Viennese-style competitions among themselves.

In some of these competitions, men and women competed together. In others, they were asked to compete separately, Victorian attitudes toward the mingling of the sexes being what they were. When the International Skating Union (ISU) created the world championship in 1896, its organizers expected only men to enter.

Women, they must have thought, didn't have what it took to formally compete. They were quite wrong.

In 1902, an English ice skater named Madge Syers registered for the championship. The rules didn't explicitly forbid women from competing, so with a begrudging shrug, the powers that be allowed her to take to the ice. Of the four competing skaters that year, Syers took second. Her success must have stung, because at its next meeting, ISU officially banned women from competing at the world championships. The official reason was that a judge might "judge a girl to whom he was attached."

Two years later, at Britain's urging, the ISU allowed women to compete by creating a "ladies" category, though the winner of the women's title wasn't considered a proper world champion until the rules were amended again in 1924. Even in the first part of the twentieth century, the separation of men and women into different categories exasperated some of the world's most prominent skaters. One of the most famous, Charlotte Oelschlägel, took aim at the policy in the introduction to her 1916 booklet on how to skate:

> This little book is . . . intended as much for women as for men. There are no physical reasons why women should not skate quite as well as men. Skating is a matter of balance and grace, not strength. . . . Up to within a few years ago, the figure skating championships of Europe were open to both men and women on equal terms.

And then the kicker: "Perhaps the fact that women excelled in grace was partially responsible for the separation of the sexes in these championships."

Despite such gendered segregation, ice skating allowed women to compete professionally at a time when few other sports were open to women at all. In figure skating, at least, the creation of the ladies' category didn't relegate women to second-class status as the category did in many other sports. "The ladies did the same compulsory figures as the men, and because there were no required elements in free skating, program content was neither dictated nor limited by gender," writes James R. Hines in *Figure Skating in the Formative Years: Singles, Pairs, and the Expanding Role of Women.*

In figure skating, men and women included the same elements on the very same ice. So elements like music choice, choreography, and costuming became ways to distinguish the masculine skating styles from the feminine. Such differences were thrown into sharp contrast during the 1968 Olympics when the American Peggy Fleming won the gold. On the men's side, the U.S. skater Tim Wood took the silver. Whereas Fleming was balletic and feminine, using styles reminiscent of Haines's, Wood skated in a style that harked back to nineteenth-century England, never raising his arms above his head. He kept his upper body mostly rigid, using his feet to form figures on the ice.

In *Artistic Impressions: Figure Skating, Masculinity, and the Limits of Sport*, Mary Louise Adams writes that the style that had come to characterize men's skating was a deliberate attempt to counter the sport's feminine associations. The American skater Scott Hamilton asked a reporter at a 1982 world championships press conference, "When are you guys going to start treating us like athletes?" In response the reporter said, "When you guys stop dressing like Liberace." After that, Hamilton wore speed skating suits on the ice, seemingly drawing the line between art and sport with attire, if not only style and movement.

In more recent years, skaters have pushed back on the notion that skating has to be gendered. In 2022, the U.S. skater Timothy LeDuc became the first openly nonbinary athlete to participate in a Winter Olympics. LeDuc's appearance at the Games coincided with (though wasn't necessarily the cause of) the International Olympic Committee ending its reliance on testosterone levels to decide which athletes are eligible to compete in male or female events.

Ice has inspired humans for thousands of years to take risks and push against social norms. It makes sense that one of the most popular sports to ever take place on ice would be the one to push the boundaries of gender.

As figure skating evolved into its own competitive sport in the U.S., ice spectacles of another kind were taking shape. In the winter of 1865, the nation was embroiled in civil war. But not everything felt grim. This was the year that traveling carnivals on ice began touring in earnest on a circuit that took them from Philadelphia to Chicago and then up to Montreal and down to New York, then round and round again. With each tour, the carnival attracted more people desperate to escape the despairing reality of war-torn life. They used the carnivals as excuses to dress to the nines, a choice frequently pooh-poohed by traditionalists. Wrote one stodgy reporter at the time: "Fancy Dress Entertainments and Masquerades" should remain distinct. "A mingling of the two . . . is against all tradition."

Such criticism did little to dissuade Americans from having fun at the rinks. Americans were going wild for ice sports, leading entrepreneurs of the 1870s to wonder, What if there was a way to produce these sports and sell tickets all year-round? There was only one problem: ice melts. Come summer, the rinks became barns for boat building, which was a lot less entertaining to watch.

Then there was the problem of air. The rinks were freezing cold and poorly ventilated. Cigar and pipe smoke caught in the rafters and wafted down, choking spectators and skaters. Clearly, better ventilated, more permanent structures needed to be built. But how could anyone justify the expense of a permanent structure that was used only four months of the year?

The answer seemed simple, even if the means to making it a reality was not: ice rinks needed mechanically made ice, the kind already emerging from ice plants across the country. The tricky part was that the ice needed to be long, thin, and smooth, rather than in blocks.

The idea had already been floated at least thirty years earlier, in 1841, at a rink in London called the Alpine Room. Its walls were painted with alpine scenery and a Swiss cottage. In summer, skaters wore their regular ice skates, but the surface on which they glided wasn't made of ice. The wooden floor was coated with hog fat and soda. Skaters reported that moving through it felt like cutting through hard cheese. Falls were to be avoided at all costs.

Experiments with actual water—and then with ammonia—were conducted in the years that followed, but they failed. Not until 1876 was there a breakthrough. John Gamgee opened the Glaciarium, a small, short-lived ice rink in London whose ice was created via a coolant that was distributed through coiled pipes below the floor. Based on John Gorrie's patent, the contraption involved pouring water on the floor so that the pipes below would freeze the water into ice in a matter of days. The invention technically worked, but the rink was too small to accommodate the number of skaters who sought to use it. It closed in just over a year.

The United States opened its own year-round ice rink a few years later, a rink that would eventually become the most famous multipurpose indoor arena in the world. And it all started with one of the most famous (and most problematic) showmen in American history.

On July 13, 1865, a siren rang out through New York City's financial district. Half a dozen horses pulling carriages carrying leather fire buckets galloped around the corner of Broadway and Ann Street en route to a horrific sight. P. T. Barnum's American Museum was engulfed in flames. Thousands of spectators gathered to watch aghast as the five-story wooden building collapsed, destroying curiosities like the Feejee mermaid and crushing the live animals that Barnum kept in cages and aquariums. The museum had already gained a reputation for human trafficking and cruel voyeurism, a reputation built on displays such as that of conjoined twins Chang and Eng and Feodor Jeftichew, who Barnum advertised as "Jo-Jo the Dog Faced Boy." Thankfully, the showman's human spectacles, at least, made it out alive.

Barnum reopened the museum several blocks uptown, but just three years later, that museum burned, too. What he needed, the showman thought, was a commercial structure less susceptible to fire but big enough to house thousands of spectators—a rather difficult combination of needs to meet in the nineteenth century.

He eventually found what he was looking for in an abandoned train depot on East Twenty-Sixth Street and Madison Avenue. The roofless building had four towers, one at each corner, to which Barnum attached a large canvas tent. He encircled the perimeter with spectator seats and threw down a wooden floor. He called the building the Roman Hippodrome and, according to advertisements placed in newspapers across the country, began staging a series of acts based on "sports of Ancient Greece and Rome, and Historical Pageants in the world." The ads also boasted that the show was "ten times the largest show on earth," an exaggeration to be sure, but not by much. The space held ten thousand people, an audience rarely seen anywhere else.

The show consisted of chariot races around a dirt track and, later, a parade of elephants. The animals were the biggest draws, so Barnum made them the stars of the show. Within a year he rebranded the Hippodrome as the "Greatest Show on Earth," and it became America's most popular circus.

Barnum soon learned he could make more money on the road, so he leased the Hippodrome to another entertainer, Patrick Gilmore. Gilmore was an Irish composer and bandleader who fought for the Union Army in the Civil War and was most famous at the time for penning the lyrics to "When Johnny Comes Marching Home." By the time Gilmore was offered a lease for the Hippodrome, the Civil War had been over for ten years, and he was living in Boston, where he had staged two enormous music festivals in celebration of national reunification. Johann "the Waltz King" Strauss made his only American appearance at one of these festivals. At the second, Gilmore became the first bandleader to feature a saxophone.

Gilmore transformed the Hippodrome into a concert venue, but he grew bored. After years of concert planning, he was ready for a challenge. For something that would truly transfix the American public. Inspiration struck during the winter of 1878 as he sat in Central Park watching the ice skaters. He would erect an indoor skating rink that could be used throughout the year.

The only successful year-round rink he likely knew of was Gamgee's Glaciarium, and his knowledge of even that is up for debate. He collaborated with several American and European engineers on the ice-making mechanics of an indoor rink, and within a year he had finished building a system comprising over a mile of wrought-iron pipes that transported liquid ammonia brine through a layered floor, cooling it to water's freezing point. Four inches of water were sprayed on the surface, and once it froze, the floor held more than 16,000 square feet of flat, sparkling ice.

Inspired, perhaps, by Barnum's spectacles, Gilmore planned an outrageous event to reveal his great indoor ice rink. He hired skaters from touring carnivals to perform in bright, colorful costumes, and encouraged spectators to bring their own skates to join in the fun. On opening night, the space was sold out, with spectators eager to leave their seats for the awe-inspiring ice before them. The Hippodrome had been completely transformed. Here was an event space inspired by fire, but it now flourished because of ice.

A newspaper article raved about the "thousands of gas jets, hundreds of colored lights, and the flashes of numerous calcium lights of different colors" that filled the space. Meanwhile, a band played Strauss's waltzes along with Gilmore's own compositions. The skating thrilled the public, but the real draw was the ice rink itself. The evening's MC, Robert Gibson, one of the most popular ice skaters in the world, stopped the show midway to praise the engineers for their ingenuity: "We who have enjoyed the privilege of skating upon the first artificial ice rink ever produced in America and of standing upon the largest cake of ice ever made by man . . . extend to you our heartiest congratulations on the magnificent success you have achieved."

Gilmore renamed his event space Gilmore's Garden, and its rink inspired other mechanically made ice rinks to open across the country in Detroit and Pittsburgh, and eventually in Europe. A few years later, he sold the Garden to another developer, who renamed the space for its address: Madison Square Garden.

I n 1905 another event space, the 5,200-seat Hippodrome Theatre, was built in New York City in the middle of the theater district. The space featured state-of-the-art theatrical technology, including a rising glass water tank and a mobile ice-skating rink. The theater's producer, Charles Dillingham, went to Europe, then still the home of "fancy skating," to find a skater to star in *Hip-Hip Hooray*, his new stage show set in a ski lodge in winter. He found his leading lady in Berlin, a young Charlotte Oelschlägel performing in Leo Bartuschek's ice ballet *Flirting at St. Moritz.*

Hip-Hip Hooray opened on September 30, 1915. The show's comedy skits and musical acts were punctuated with performances by dancing horses. Nothing compared with the third act, however. The curtains rose on a skating tableau set on an expanse of glittering ice. Snow fell from the rafters while skaters waltzed and sang and fell in love. Broadway had never seen anything like it. The show ran for a record 425 performances in three hundred days.

The show's spectacular set design and special effects were what initially attracted audiences, but the show's star enthralled them. Wearing white boots and daringly short skirts that barely reached her knees, Oelschlägel spiraled on the ice with one leg straight up behind her and her body angled downward so that her hair swept the ice. The move became known as the Charlotte Stop. Michelle Kwan would include it nearly a century later in her long program during the 1999 season.

It didn't take long for Hollywood to come calling. Oelschlägel dropped her last name, going only by Charlotte, and became the first skater to appear in cinema as the protagonist in the 1916 six-part series *The Frozen Warning.* Charlotte plays a college student from Vassar who learns of a plot to steal government secrets. She warns officials of the impending danger by carving the word "spies" into the ice during a skating party.

During the 1920s and '30s, Charlotte performed throughout Europe and North America. She married her skating partner, Curt Neumann, in 1925, and together they're credited with devising the backward outside death spiral, a spin in which one partner lowers the other close to the ice. The lowered partner arches backward and lifts a foot while holding the upright partner's hand, while both balance on the outside edge of their skates. The couple returned to Berlin in 1939 to attend Charlotte's mother's funeral and were trapped for the duration of the war. The couple lived out the rest of their lives in West Berlin. In 1985, Charlotte was elected to the World Figure Skating Hall of Fame.

Around the time that Charlotte left for Berlin, a Norwegian Olympic skater named Sonja Henie departed for the United States. Henie idolized Charlotte and, like her idol, wanted to star in skating films and Broadway shows. In 1936, Henie caught the eye of a Hollywood producer who recognized in her skating a kind of athleticism never before seen on the ice in America. He cast her in the film *One in a Million*, shot at Madison Square Garden. She played Greta Muller, a young Swiss skater who wants to win an Olympic gold medal. Henie's second film, *Thin Ice*, took a twist on the fairy tale of Prince Charming, set to classical music. Her dancing evoked the sinuous, balletic moves of Jackson Haines, and the film was a box office hit. She went on to release nine more skating films over the next ten years, including *The Countess of Monte Cristo*. Her films inspired audiences to attend her live shows, and her shows, in turn, inspired ticket sales for her films. She became one of the most well-known celebrities in America, which is especially notable in light of the fact that no American skater had yet won an Olympic gold medal or been a world champion.

Henie's work further intensified the country's interest in ice performances, and in 1940 the growing popularity of these spectacles inspired a Pittsburgh rink owner named John H. Harris to create a skating show between periods at hockey games. The ten-minute show, which at first featured just a couple of figure skaters performing routines based on Henie's work, became so popular that he swelled it into a twelve-skater revue. He dreamed of an ice carnival, like those in nineteenth-century Montreal and St. Paul, but with modern costumes and music. He met with several arena owners in Hershey, Pennsylvania, and together they created a show that would tour their respective cities. They called the show the *Ice Capades*.

The *Ice Capades* employed seventy-five performers, including comedians, clowns, jugglers, and dozens of dancing skaters he dubbed the Ice Ca-pets. As Bryan Curtis writes in "The Ice Capades: Requiem for the Ice Carnival," a "1942 souvenir program lists the pets' vital statistics with Playboy-like precision." Their "average height 5 ft. 3 ½ inches; average weight, 116 pounds. There are 21 blondes, 24 brunettes; 2 with black tresses and one auburn-haired in the group."

The early shows borrowed heavily from vaudeville. One minute, an Olympic skater is gliding and spinning to Beethoven, and the next, the Ice Ca-pets are kicking their skates in unison to "Pan Ameri-conga." Elegant waltzes bled into juggling acts. Olympic champions mingled with men in elephant costumes.

Audiences loved all of it. In the *Pittsburgh Press*, the critic Kaspar Monahan criticized the skaters' suggestive postures but concluded that "on steel runners these vulgarisms become something approaching art." In a world embroiled in war, the *Ice Capades* offered spectacle, escapism, and a dash of something like patriotism. What, after all, is more American than sexy kitsch? As Deborah Brandt, a former Ice Ca-pet, put it, "This was a Las Vegas revue on ice for families."

The *Ice Capades* continued to entertain audiences for several more decades, but by the 1980s, the show's novelty had worn off. In an era of video games and Saturday-morning cartoons, indoor ice-skating revues no longer inspired awe and wonder. Producers did their best to compete for families' attention by adding characters that would appeal to kids. A 1983 souvenir book illustrates the transition on its cover, where a dozen women skaters pose in scant, sparkling two-piece leotards and feathery hats, a costume design somewhere between Busby Berkeley and that decade's top Vegas shows. All that suggestive glitz is undercut, however, by the appearance of two skaters in Smurf costumes peeking out of the cover's upper-right-hand corner. Beneath them are the words "And SMURFS, too!"

By the 1990s, ownership of the *Ice Capades* had exchanged hands no less than six times, and by 2000, the show disappeared completely. America's obsession with ice carnivals had come to an end.

CHAPTER 10

Cork-Bungs, Brooms, and Zambonis

A hockey player on a frozen lake (1940s)

One of the first films made in America was of a hockey game. The twenty-five-second shot was taken on Crystal Lake in West Orange, New Jersey, in 1898 by the moving-picture pioneer Thomas Edison's film crew. The camera captures twelve hockey players in white and dark sweaters smacking a puck across the ice. The men worked for the Canadian railroad and loved to skate in their spare time. They formed several amateur hockey teams that played each other throughout the winter at depot stops whenever time allowed.

In the film, the players' dark clothing stands in stark contrast to the glistening white ice, creating a chiaroscuro effect that looks quite modern. Edison may have filmed the game for more than aesthetic reasons, however. In the 1890s, hockey was still a rare sight in the United States. But it was about to catch on in a big way.

The history of hockey is deeply entwined with the history of ice. The sport arrived in America just as figure skating was reaching a peak in popularity and at the cusp of the switch from natural ice rinks to those made with mechanically created ice. It also arrived during a severe economic depression that brought the Gilded Age to an end. Rink owners were looking for ways to earn more money, and hockey proved to be their golden ticket to riches beyond imagination.

During the 1870s and '80s, the United States experienced unprecedented economic growth, but such growth was dependent on the high price of commodities sold in international markets. The 1890s witnessed a global plunge in the price of goods of all kinds, and especially wheat—one of the country's biggest exports. At the same time, the United States was experiencing an oversupply of silver, which devalued the metal; and the railroad industry, by now only a couple of decades old, was forming what might be called today a "bubble," as more tracks were being laid than used. All this led to the Panic of 1893, the worst economic fallout the country had ever seen, surpassed only by the Great Depression thirty years later.

Unemployment rates soared to 25 percent. To feed their families, out-of-work men and women earned money by chopping wood and sewing by hand. Some women turned to prostitution. Everywhere, theaters and music halls closed, but the popularity of ice skating only increased. Ice skating provided a relatively inexpensive distraction for families who had few other options. Rink owners went searching for novel ways to keep their skaters coming back. They discovered a stick-and-ball game that had become popular on frozen ponds near train depots and wondered how they might profit from it.

This strange game was new to America, but not to the world. The first recorded instance of such a game, "chamiare" or "shinty," appeared in the Firth of Forth in Scotland in 1608, during the second half of the Little Ice Age and one of the coldest and stormiest winters in history. A similar game called bandy emerged in England and spread throughout the British Isles. According to the nineteenth-century historian Charles Goodman Tebbutt, men had probably been playing bandy on ice since the mid-1700s. "Concurrently with skating races, bandy matches have long been held in the Fens," wrote Tebbutt in 1892.

The word "hockey" possibly made its first appearance in a 1773 book entitled *Juvenile Sports and Pastimes*. Its author, Richard Johnson, describes a game called "hockey" wherein players hit a "cork-bung" (a beer-barrel plug) around a plane of ice. Cork-bungs were big in hockey's early days. Two decades later, in 1797, the artist Joseph Le Petit engraved a picture of hockey players knocking around a cork-bung on a London rink. The fact that the word "hockey" emerged at the same time that cork-bungs were used as pucks has led some historians to suggest that "hockey" may come from "hock ale," a beer brewed for the festivals at Hocktide that came in barrels plugged with cork-bungs. ("Hockey" was also used by London locals as a synonym for "drunk.")

Other early stick-and-ball games included *hurley* in Ireland, *kolf* in Holland, and *knattleikr* in Iceland. All were played on ice. In the Americas, First Nation communities played stick games called *tooadijik* and *wolchamadijik* on ice. Those likely merged with games brought over by English and European settlers to create something close to what we think of as hockey today.

Some of the earliest evidence of ice hockey in North America was recorded by British explorers. In 2002, researchers discovered two letters written by Sir John Franklin in 1825 during one of his attempts to find the Northwest Passage. He and his men were wintering in the Northwest Territories and kept themselves entertained with what the locals called "hockey on ice." Those amusements came to an end on October 20, 1825, with the arrival of the first major snowstorm of the season. In his journal, Franklin writes that the storm "continued without intermission for 36 hours." In another letter, Franklin notes that he and his men kept themselves "in good humour, health, and spirits by an agreeable variety of useful occupation and amusement. Till the snow fell the game of hockey played on the ice was the morning's sport."

The first hockey match in its modern-day form is said to have taken place on March 3, 1875, when a natural-ice skating rink in Montreal, Canada, made a significant change to how the town's favorite ice game was played. The change came after the rink advertised that two skating teams would compete to knock a rubber ball into opposing nets using long sticks—a description that made some potential spectators wary. As a journalist for the *Montreal Gazette* put it, people were concerned that "accidents were likely to occur through the ball flying about in too lively a manner." The rink issued a statement to appease the worriers saying that, instead of with a ball, "the game will be played with a flat circular piece of wood, thus preventing all danger of its leaving the surface of the ice." And lo, the hockey puck was born.

As luck would have it, this particular match also inaugurated another time-honored aspect of hockey: hand-to-hand (and stick-to-stick) combat. Deep in the game, a couple of young boys in skates, too tantalized by the ice to stay in their seats, took to the rink and began skating among the players. According to an account by the *Daily Witness*, the intrusion resulted in an "unfortunate disagreement." One of the boys was "struck across the head, and the man who did so was called to account, a regular fight taking place in which a bench was broken and other damage caused." Another report suggests that the fight resulted not from a couple of young boys but when the local skating club—already feeling territorial about their ice—arrived and forced the game to end early. A newspaper from Kingston, Ontario, reported, "Shins and heads were battered, benches smashed and the lady spectators fled in confusion."

The sport made its way to the United States via the railroad. Once it arrived, it caught on quickly among ice skaters looking for other ways to amuse themselves on ice. The year before Edison's camera crew filmed the hockey game in New Jersey, hockey's first guidebook,

Spalding's Ice Hockey and Ice Polo Guide, made its way into the hands of skaters eager to learn the sport. The book observed that "three winters ago Chicago, Minneapolis and Detroit were about the only scenes of the game's activity, but last winter where ice could be found . . . ice hockey was being played."

As hockey spread throughout the country, rink owners came to see it as a way to draw spectators for an evening and charge admission. They converted their rinks into dual-purpose spaces by building bench seats around the ice and covering the ice with ramshackle roofs. Some added thin wooden walls to trap heat and keep attendees warm—innovations that laid the groundwork for the state-of-the-art multipurpose arenas that host hockey matches and other entertainments today. They limited ice skating to certain evenings so that hockey could be played on the other nights. A hockey rink in Pittsburgh, Pennsylvania, called the Winter Garden advertised its variety of offerings by publishing a song entitled "Winter Garden Glide." "Listen jolly skaters to the band," went the tune. "It's grand." The sheet music's cover depicts eight images of twirling skaters circling a photo of the ice rink. Below them, two hockey players break the frame. The players have nothing to do with the song but everything to do with the dual nature of the rink.

The novelty of the sport—and, well, its unpredictable violence— attracted thousands of new fans, whose ticket purchases provided rink owners with the funds to convert their natural-ice rinks into those with mechanical ice, making it possible for the sport to be played (and money to be made) year-round. Indoor, multipurpose skating rinks built on mechanical ice went up across the country. Among them was Pittsburgh's Casino, operated by the Schenley Park Amusement Company. The establishment contained luxury boxes, a theater, an indoor ice rink, and seats to hold an audience of twelve thousand. A report called the Casino the "handsomest amusement building in the United States," and the *Pittsburgh Commercial Gazette* referred to it as "the magnificent palace of pleasure."

The Casino opened during an unusually warm May in 1895 to thousands of wide-eyed spectators who'd never seen a rink of ice during such temperate weather. "Skating on ice is a pastime that perhaps no one in this city ever before indulged in at this period of year," claimed an article in the *Pittsburgh Press*. And in the *Pittsburgh Post*, a journalist marveled at how "hundreds of persons took advantage of the rare opportunity to skate on ice in hot weather."

If to see ice in late spring was a delightful surprise among Pittsburgians, then to witness a game of hockey for the first time was an outright astonishment. In December of that year, a team of Queens University polo players met a group of local skaters at the Casino for a hockey exhibition. Approximately three thousand people showed up to watch and learn how hockey should be played. An article in the next day's *Press* claimed, "This game has never before been seen in Pittsburgh, and it was a revelation." The writer went on: the locals "didn't know just what to do with that little flat 'puck' used in hockey. They didn't know whether it was good to eat or whether it was a holiday toy."

For the next year, the Casino sold out every hockey match it hosted until December 1896, when the ice rink's cooling system sprung a leak. It caused an explosion that started a fire, and in just two hours, the "palace of pleasure" burned to the ground.

———————

By the 1920s, World War I had ended, and ordinary Americans had more leisure time and ample disposable income. At the same time, Americans were becoming enamored of so-called manly sports. Boxing became legal in the United States, and ice hockey, with its semi-legalized fighting during games, gained an even larger fandom. In 1924, the Boston Bruins joined the National Hockey League, and by 1931, the NHL was boasting eight franchises in Canada and the United States, all housed in newly constructed permanent arenas equipped with their own ice plants. The universal availability of mechanically made ice at this point eliminated the uncertainty of weather from the scheduling of games, and hockey's season doubled in length from twenty-two games to forty-four.

These arenas also provided comfort unlike anything experienced before at a skating rink, with heating and air-conditioning, bathrooms and food concessions. Some even offered padded seats. Such luxury was so unusual it led to a lawsuit in 1927 between T. P. Gorman, manager of the New York Americans franchise, and Tex Rickard, president of Madison Square Garden. Gorman carped that Rickard kept the temperature inside the Garden too high (seventy degrees) and his players were overheating on the ice. Even the ice itself, he argued, was melting. The *New York Times* reported on the suit, suggesting that Gorman kept the Garden warm so that "women spectators in evening gowns sat with their wraps removed." (Despite hockey's reputation as a rough sport, social mores of the era demanded that spectators wear their fanciest clothes.)

As arenas grew bigger, owners experimented with switching from skating to hockey in the same evening. From a financial point of view, this was a grand idea. But the damage all that skating did to the ice caused numerous accidents. Ice skating is hard on ice. Skates nick and chip and scrape. Ice cracks from the weight of too many skaters. The process of repairing ice was long and labor intensive, and there was never enough time before games started to fill in the cracks. When hockey players took to the ice, they'd arrive on a sheet almost too dinged up to play their game.

This problem was solved in 1949 by Frank Zamboni, owner of an ice plant in Southern California. Ten years earlier, Zamboni had noticed that the popularity of artificial refrigeration was leading to a drop in demand for the block ice he made at the plant. He called a cousin, and the two of them reconfigured the plant to form a rink they called Iceland. At 20,000 square feet, it was one of the largest mechanically created ice rinks in the world. It hosted more than eight hundred skaters per day, all of whom took a toll on the ice. To repair the damage, Zamboni and his team pulled a blade by tractor to shave off the rink's top layer. Four employees would follow the tractor with shovels, scooping away the shavings. The whole process took over an hour and a half.

Zamboni tinkered away on a machine that could complete the process faster, and in 1949, he introduced his invention to the world. Named after himself, the Zamboni could complete the entire ice-resurfacing process in fifteen minutes. That first machine was built on a Jeep frame. A wooden box in the back held the ice shavings. As it moved along the ice, water dumped from the back to wash and replenish the ice, leaving a flat, fresh surface.

Zamboni improved the model a year later, when figure skater Sonja Henie observed the first one in use and asked that he build one just for her. He patented the improved version in 1953, and one year later, the machine made its first appearance at an NHL game. Six years later, it was scraping ice at the Winter Olympic Games.

By 1956, Zamboni had sold thirty-two machines. Today, more than ten thousand have been sold. The Zamboni has become an iconic machine of the ice, nearly as recognizable to fans of ice sports as the hockey puck. There's something inherently amusing about its boxlike shape and slow-moving plod across the ice (its speed maxes out at nine miles per hour). In 2009, John Branch of the *New York Times* wrote, "Zamboni may be the most famous name on ice, a pop-culture icon more recognized than any of the four remaining National Hockey League playoff teams." David Letterman drove a Zamboni during a segment on *Late Night*. When the machine appeared on camera, it received more cheers and laughs than any of the jokes leading up to it. Parker Brothers turned the machine into a player piece for a hockey-themed Monopoly game. The cartoonist Charles Schulz, who famously loved hockey, once wrote a comic strip where Charlie Brown remarks, "There are three things in life people like to stare at: a flowing stream, a crackling fire and a Zamboni clearing the ice." The machine appeared time and again in *Peanuts*. In a strip from 1994, Woodstock drives a bird-sized Zamboni over a plane of ice, making it so slippery that three of the Peanuts gang are sent airborne after stepping on it. Snoopy watches the scene, thinking, "That Zamboni makes good ice . . ." In another strip, Charlie Brown chastises Snoopy for driving his dog-sized Zamboni in the house.

The Zamboni helped revolutionize ice sports because it made the ice safer and more consistent in thickness and smoothness. That's important in hockey, because hockey ice isn't made to the same specifications as ice for speed skating or figure skating. It has to be stronger because of the players' sharp turns and short, wide blades, which gouge the sheet. According to the NHL, hockey ice should be kept to at least three-quarters of an inch so that it's easier to repair.

Today, the Zamboni does most of that repair work, but hands-on labor by actual humans is still required, especially when the ice gets too thin in places. There's no easy way to patch ice past a certain thinning point, even with a Zamboni. Workers use whatever's available to get the patching done. It's not unusual to find in an ice workers' tool belt a few spoons for scooping out goalpost holes or a turkey baster to wet surfaces. The Zamboni needs repairs, too. Broken hockey sticks are good for tapping slivers of shaved ice off its blades.

Hockey may have evolved into a sport with pro leagues around the globe and its own iconic vehicle, but when it comes to repairs, it still evokes the scrappy game played by nineteenth-century railroad workers and families fallen on hard times.

The Nutmeg Curling Club sits in a wooded park in the middle of Bridgeport, Connecticut. Its vintage-looking sign brings to mind the early twentieth century, when professional ice sports first began to capture the attention of the American public. When I arrived, I was brought inside to the "warm room," a carpeted space in the back of the club that looks and feels like what it's called. With its wood paneling and big brown armchairs, the room suggested my grandparents' house, the warmest place I ever knew. Along the walls were shelves spilling forth with trophies and medals and plaques won by the local league. Most of the awards were dated from the 1960s and '70s. There was a bar in the back, and though it was only 11 a.m., a bartender was already on duty. One of the club managers asked if I wanted a drink, and while I was tempted to ask for a beer to settle my nerves, I told her maybe after my first lesson. From what I'd seen of curling on TV, I was going to need a clear head.

One wall was made completely of windows, which looked out over the ice. I watched as a man wearing a backpack sprayer walked backward on the sheet, shaking the nozzle to and fro to spray the ice evenly with water. This, I learned, was the famous "pebbling" procedure, wherein beads of ice are formed to create the distinctive textured surface required for curling. The pebbled texture is crucial to the sport—even though, as I was about to learn, the experts aren't quite sure why.

Curling is a game played with stones and brooms, the only sport I know of that requires constant brushing. It's played on a long strip of ice called a sheet that measures 45 meters in length and as many as 5 meters wide. At each end of the sheet are circles that look like targets; these are called houses. Players slide and spin a rock of granite called a curling stone down the sheet, aiming for the house on the opposite end. To help the stone arrive at its target, players frantically sweep the ice around the stone as it slides, while the team's captain (called a skip) shouts "Sweep!" to keep them moving. The sweeping keeps the stone sliding and can even change the direction of its path, as necessary.

As with most ice sports, the origin of curling is murky. It likely began in sixteenth-century Scotland, where people flung stones over frozen lochs. Some historians argue that it was actually the Dutch who invented the sport, pointing to mid-sixteenth-century paintings such as Pieter Bruegel the Elder's *Hunters in the Snow* as evidence. The artwork depicts in the background a pond with figures that look like they might be playing a game of curling. Immigrants from Scotland or Holland probably spread the sport to North America, where, like hockey, it caught on first in Canada and then made its way south. The first U.S. curling club opened in Pontiac, Michigan, in 1828.

Curling didn't appear in the Olympics, however, until 1924, at the games in Chamonix, France—the first Olympic Games to feature winter sports. At the 2022 Olympics, neither the U.S. men's nor women's curling teams won medals, but they gained some high-profile fans, including the American actor Laurence Tureaud, better known as Mr. T., who tweeted the following in the weeks leading up to the 2022 Games: "I am pumped! I am excited and I am ready! You know why? It's time for The Winter Olympics! And you know what that means! My Favorite! The U.S.A Curling team is back! I am sorry, the Defending Olympic Champions Curling Team!"

The rules of the game aren't hard to grasp, but playing the sport well requires balance and strength. Curling stones weigh between 38 and 54 pounds and are made of granite harvested off the west coast of Scotland. They're too heavy to fling like a skipping rock. Curlers slide them down the sheet using an underhanded motion reminiscent of that used in bowling. The teammates begin sweeping the ice about a third of the way down the sheet. If they sweep hard and fast enough, the stone will change its trajectory toward the target. So where does the "curling" part come in? By adding a little spin to their throw, curlers can make the stone "curl" along its path to block an opponent's stone or even knock it out of the way. The tiniest rotation can deflect the path of a curling stone by as much as five feet. What's weird is that the stone doesn't curl the way you'd expect.

Despite the sport having existed for centuries, no one, not even scientists who study the physics of ice and motion, knows why curling stones move the way they do. Many scientists believe their strange motion has something to do with the ice itself. Curling ice must be perfectly flat—far flatter than ice found in a hockey rink—before it's pebbled. Many variables come into play during the pebbling process, which can affect how the pebbles are created. This includes how quickly the person spraying the ice is walking, how quickly the spray nozzle is shaken back and forth, the temperature of the water, the distance between the sprayer and the ice, and the air temperature of the rink itself. If just one variable is not accounted for, the pebbling won't work.

Once pebbled, a Zamboni is used in some clubs to shave the tops off the droplets, leaving them smooth. This also leaves gaps between the droplets where air can flow freely, helping the stone glide faster and curl in a more precise manner. But experiments prove that none of this accounts for the weird physics at play on the ice.

In a popular YouTube series called Smarter Every Day, the host, Destin, demonstrates this unusual curling phenomenon using a drinking glass. He slides the glass upside down on a table, adding a spin as it leaves his hand. The glass follows a curved path across the surface, arcing to the left when he spins the glass clockwise, and to the right when he spins the glass counterclockwise. The directions in which the glass curves can be replicated in almost any context on earth, because most objects glide in the opposite direction in which they are spun. This happens because the spinning motion generates friction, which slows the object's rotation until it comes to a total stop. But curling stones behave in the opposite way. A clockwise spin causes the stone to arc to the right, and a counterclockwise spin moves it to the left.

For years scientists assumed this odd behavior was the result of something called asymmetrical friction, an effect produced by extra pressure at the front of the stone, which melts the ice and creates a thin film of water that makes the surface beneath the stone more slippery. This theory was widely accepted until 2012, when a research team at Uppsala University in Sweden actually tested it. They discovered that curling stones rotate surprisingly slowly, completing just two turns before coming to a stop. Such slow spinning can't produce a force strong enough to generate asymmetrical friction.

What's even stranger is that curling stones curl only when players apply the gentlest of motions. Try to spin a curling stone too fast, and it won't curl at all. This isn't how anything else in the known universe behaves on ice.

Puzzled, the researchers employed an electron microscope to examine the ice under a curling stone. They discovered that the front edge of the object leaves tiny scratches on the sheet in the direction of the rotation, acting as guides for the rest of the stone to continue its spin. In other words, the scratches are "steering" the stone in the desired direction. The scientists were able to replicate the phenomenon themselves by scratching the ice in various ways.

But what this new model didn't account for was that the arc of a stone's curl happens over the course of about three feet—a distance longer than what the scratches could create on their own. More recently, a third theory has emerged, something called the pivot-slide model. In this model, the stone creates what's referred to as stick-slip friction whenever it encounters a pebble. The stone catches on the pebble, pushing it back, and because ice moves at the microscopic level like elastic, the pebbles snap back into place, further pushing the stone on its path in the process. The sheer number of pebbles on the ice could create enough push to curl the stone along its arc, but the experiment hasn't been satisfactorily replicated to adequately prove the hypothesis.

The questions of how or why curling works remained unanswered as I took to the ice for my first curling lesson. The air temperature above the indoor rink was about forty degrees. Dressed in a windbreaker and a ball cap to keep warm, I put my left foot in a glider (a tool to help newbies like me glide across the ice) and my right hand on the stone. (Professional curlers wear specially made, mismatched shoes on the ice: one with a smooth sole made of Teflon, called a slider; and another with a rough sole, called a gripper. My abilities were not yet at the slider level.)

My coach, who was standing halfway down the sheet, gave a signal to throw, so I pushed off, using the glider under my foot to guide me. I attempted to form a pose like a bowler as I glided—something I'd seen while watching the 2022 Olympic curling team—and immediately felt my foot slip out from under me. The stone went sailing, and I fell onto my side. The handful of spectators burst out laughing, and feeling the pebbled ice beneath my palms, I laughed along with them.

Later, I spoke with the club's president, Alexis Boccanfuso. I asked what drew her to the sport, and she mentioned its social aspect—the Nutmeg Club competes in "bonspiels," or matches, with other regional clubs throughout the year—but she also loves thinking about the ice. Curling ice, she said, "changes constantly throughout the game" as friction from stones, brooms, and the players' feet subtly reshapes it. That sheet of ice, she continued, is "the single most important equipment [curling players] have. It's critical to know what the ice is doing" during the whole match to play well. When I asked her what could have contributed to my own poor performance on the ice, Boccanfuso laughed and said that balance and technique come with practice. The most important thing a new curler could do, she said, was learn to "read the ice." I asked her how one does such a thing, and she said that ability, too, comes with experience. "Ice," she said, "is like a living being." It reveals who it is with time.

CHAPTER 11

The Need for Speed

Larry Jensen competes in the juvenile boys' quarterfinal event
during the Silver Skates Derbies (1954).

When asked whether ghosts are real, Stephen King replied, "Reality is slippery. People slide on it and break their bones all the time. And if the ice is thin you can fall through." King's use of ice as metaphor for the supernatural was more apt than he probably realized. The physical properties of ice are strange, spooky even—especially those related to slipperiness. People can't skate, for example, on silver or granite, even though, like ice, both are strong and smooth. But why not? What makes ice in particular so slippery?

No one knows for certain, not even the scientists who study the phenomenon.

In 1939, the scientists Frank Philip Bowden and T. P. Hughes published a paper that put forth one explanation: that ice is slippery because of friction. The skater's blade rubs the frozen surface, so their argument goes, creating heat and melting the top layer of ice, making it slippery. But anyone who's ever stepped out their door on a blustery winter day and onto an icy sidewalk, only to have their foot fly out from under them, knows that this theory doesn't hold water. That fraction of a second during which the foot makes contact with the sidewalk isn't enough time to generate the heat necessary to melt the ice.

Yet, Bowden and Hughes were on to something. Friction does make a difference. Too much pressure from the skate—that is, too much friction—and the skater comes to a complete halt. Not enough pressure and the skate won't move at all. Friction, therefore, has *something* to do with slipperiness. But what?

The answer most accepted today dates to almost a century before Bowden and Hughes published their study. In 1842, the English physicist Michael Faraday proposed that ice has a "quasi-liquid" top layer that's too thin to be detected by the naked eye but thick enough to create slipperiness for a skater's blade. According to his theory, the blade creates friction instantaneously as it glides across the ice, but not by lowering the ice's temperature. It creates fiction by exerting the slightest of pressure to disrupt the water molecules across a very thin top layer of ice, changing their state from water to something called quasi-liquid. Upon this thin layer, the skater essentially hydroplanes across the ice.

Faraday called this top layer quasi-liquid because, at just a few molecules thick, it would be too thin to be classified as ordinary liquid water but too thick to be considered a solid. Faraday couldn't prove this theory in his own time, since the technology to observe microscopic particles hadn't been invented yet. But in 1987, scientists finally verified its existence using microscopes and X-rays, estimating the layer to be one thousand times smaller than a single bacterium. This discovery has led to another question, however: Since water is known to be a truly poor lubricant, how could this quasi-liquid layer reduce friction and make ice so slippery?

So far, the leading scientific answer is "it depends." In an interview for *Inside Science*, Rinse Liefferink said, "Different experiments often have different results because they are based on different conditions, and because different conditions can result in different mechanisms. It makes it a bit hard to say something general about ice friction."

In other words, slipperiness can be increased or decreased by any number of factors, including the speed and weight of the skater, the temperature of the ice and air, the degree to which existing liquid water is already on the ice surface, and the shape of the skate. When scientists ask why ice is slippery, any number of answers could be given, and none of them would necessarily be wrong.

To most of us, this question is more interesting than practical. Not many people have reason to think about the slipperiness of ice except to avoid it. But there's one group of athletes who want to know as much about slipperiness as possible: speed skaters. Finding answers to why ice is slippery can mean the difference between winning and losing a competition.

———

A t the 2022 Winter Olympics, Erin Jackson of the United States won a gold medal in the women's 500-meter event, becoming the first Black woman to win the gold for speed skating. Her win was additionally notable for being the first U.S. medal in that race since 1994. Speed skating in general is not the nation's most competitive sport. It never caught on in the United States, at least not to the degree it did in, say, the Netherlands, where the sport sells out arenas every year. Its popularity there makes sense, given that's where speed skating probably originated.

In the fourteenth century, Dutch skaters were the first to trade in their wooden blades for iron ones, which were faster and more durable than any that had come before. The upgrade led to a national explosion in hobby speed skating. The Dutch weren't the first to formalize the pastime into a sport, however. That accolade goes to the Scottish, who opened the first known speed skating club in Edinburgh sometime in the late seventeenth century. From there the sport spread to England, and the first known speed skating competition took place in the Fens on February 4, 1793.

A little more than fifty years later, in 1849, the first speed skating club in the United States opened in Philadelphia not far from the Schuylkill River. Figure skating had already taken hold in the city, drawing crowds of men and women to the frozen river all winter long. The speed skating club sparked the city's interest in the sport, and the following year, a Philadelphia-based entrepreneur named Edward W. Bushnell upgraded speed skating once again by manufacturing the first pair made of steel. These skates were lighter and stronger than those made of iron, and they didn't require the same frequent sharpening. Speed skating grew in popularity, especially throughout the Northeast and the northern Midwest.

By the end of the nineteenth century, the sport was practiced wherever winters were cold enough to freeze local lakes and rivers. Newburgh, New York, became an especially popular spot for speed skating. An 1891 city tourism pamphlet declared, "It has been said that the history of skating—that is, speed skating—in this country, if ever written, must be written at Newburgh, which is now, and, our oldest residents say, always was, 'the headquarters for fast skaters.'"

There's some truth to that claim. Several nineteenth-century skaters known for their speed called Newburgh home, including Charles Payne, a Black man who was one of the fastest skaters in the area and free before the state of New York abolished slavery. Other well-known Newburgh skaters were equally recognized for their prowess on the water, suggesting that their love of speed skating was matched only by their love of fast-moving boats. The Newburgh-born Charles June, who was famous for speed skating backward, made a living captaining barges and steamboats. He was also a champion rower. So was Timothy Donoghue Sr., an immigrant from Ireland who was crowned Newburgh's fastest skater in the 1860s. Together with a skater named Aaron Wilson, the three Newburghians skated a large swath of the Hudson River in 1872 for five hours, aided by what Donoghue called "a strong wind."

The first known woman speed skater in the Hudson Valley was a secretary named Elsie Muller-McLave. She represented the United States in the 1932 Olympics, the Games that inaugurated women's speed skating, but only as a demonstration sport. Back in the Hudson Valley, women enjoyed speed skating, but they weren't allowed to compete professionally until the mid-twentieth century. Throughout the 1940s and '50s, the New York–based Mary Lynch, Caroline Crudele, Joan Russell, and Gwendolyn Dubois won medals at national championships.

Chicago served as another groundswell for American speed skating. The Windy City hosted the first international speed skating race in 1918. The world champ was the Norwegian skater Oscar Mathisen, who came to the Windy City to defend his title. The two-day, six-race event drew thousands of fans and dozens of skaters, including the Chicagoan Bobby McLean. Mathisen had been trained by the best in the world, whereas McLean had "learned to skate on the parks of the west side," reported the *Chicago Tribune*. McLean and Mathisen both had just turned pro and, as the *Tribune* later wrote, "the smart money said the American might be able to compete at the shorter distances." On the first day of competition, McLean beat Mathisen in the 220-yard sprint as well as the one-mile and two-mile races. On day two, McLean beat him again in the 440-yard sprint. He lost the half-mile when he fell, but recovered quickly to win the three-mile race "with ridiculous ease."

Seeing an American win so handily sparked (at least temporarily) a national craze for speed skating in America. In January 1915, more than fifteen thousand people attended an outdoor race in Chicago's Garfield Park lagoon. Two years later, thirty thousand watched as

Arthur Staff won the city's first Silver Skates Derbies at Humboldt Park lagoon. The Silver Skates competition grew larger every year, inaugurating a boys' division in 1919, a women's division in 1921, and a girls' division in 1922. The Silver Skates race is still held in Chicago today.

The sport was played across the country and around the world on natural ice until the 1960 Winter Olympics, which held its games on mechanically created ice. That was also the year that women were authorized to compete in speed skating. Since then, men and women both have been setting and breaking records in the sport, mostly in Utah. In fact, more records have been set at Utah's Olympic Oval than any other place on earth, and it's not just because the record breakers are elite athletes. It's because of the Oval's ice.

———————————

Located southwest of Salt Lake City, in a small town called Kearns, the Oval is a skating facility with a 400-meter skating track and two Olympic-size ice sheets. Combined with its administrative offices and a running track that circles the ice track, the Oval boasts more than 275,000 square feet, roughly the size of four football fields.

It was built to host the long-track speed skating events for the 2002 Winter Olympics. It almost didn't open on time. A part of the roof collapsed just months before its scheduled opening. A few months later, the freeze tubes were found to have moved out of alignment. The entire floor had to be torn up and replaced so that the ice would freeze evenly across the sheet. Construction was finished and the final coat of ice was added to the track on February 12, 2001, just in time for the world single distance championship scheduled there the next month.

Since the Oval's opening, ten Olympic and nine world records have been set there—more than at any other location. Shane Truskolaski, the Oval's ice maintenance manager, is the magician behind the super-fast ice. Unlike so many others who work in the ice industry, Truskolaski didn't grow up in an ice family, though he refers to his peers at other facilities as such. "The ice industry is small, and we all kind of know each other," he told me. Truskolaski spent his childhood taking parts in plays and working with theater companies. He moved to Utah to finish his undergraduate degree in the early 2000s and found a job at the Oval's front desk. That position "didn't suit" him, he told me, so he applied to the ice maintenance department, where he became the Oval's Zamboni driver. That's when he fell in love with ice.

Talking to Truskolaski about ice felt like talking to a poet about poetry, or a painter about painting. Ice is not just his work; it's his passion. "An elite athlete can tell that they're skating on special ice," he told me. "The ice [at the Oval] is so much different than hockey or figure skating ice, definitely different than curling ice. It needs its own attention. To make our athletes succeed." He loves to be the one who gives the ice that special attention.

He explained that the key to slippery ice is making it as pure as possible. "You don't just want to have tap water that comes out of your kitchen faucet out here," he said. "It's not ideal for speed skating, because impurities [in tap water] add grit and, therefore, friction. We want to get all the impurities out of it." I asked him to give me some numbers. "So, my house's water is like three hundred parts [of impurities] per million," he said. "But the reverse-osmosis system here takes it down to about ten parts." He smiled, looking proud. "When I go to ice conferences and I tell them that, [everyone's] shocked."

Temperature is also carefully modulated at the Oval. "Our ice plant gets the temperatures of the ice super low," he told me. The sweet spot is sixteen to seventeen degrees Fahrenheit. If he takes the temperature any lower, then frost forms on the surface, which adds friction. To reduce the chance of frost, Truskolaski operates an elaborate dehumidification system that lowers the facility's dew point and keeps humidity below 30 percent.

"The problem," he told me, is that "we don't want a totally dry system either. If you have 10 percent humidity or less, the ice sublimates," meaning it destabilizes and changes from a solid to a gas. "Not to a liquid, but directly to gas. Ice is weird."

The temperature of the air above the ice also contributes to the sheet's slipperiness. During the off-season, Truskolaski keeps the air at fifty-eight to fifty-nine degrees. During competitions and Olympic trials, he raises it to the mid-sixties so that the extra heat starts to melt the top layer of ice just slightly. "I call this the air-ice interface," he told me. "And that interface, when warm enough, creates that quasi-layer, the not-completely-frozen layer that makes ice slippery." If the temperature goes any higher than that, Truskolaski said, it can affect the athletes' respiration rates and, thus, their performance. This attention to athletic performance is especially important at a facility like the Oval, which sits at an altitude of 4,675 feet. High-altitude rinks have less air density, so skaters glide through the atmosphere more easily. That's why most world records are set at high-altitude rinks, Truskolaski said.

The ice is also thicker at the Oval than at other rinks. "We try to keep our ice at around an inch and a half," he said. Beneath all that ice are wires that need protection. And when the rink is open to the public, which it is throughout the off-season, the thick ice protects the facility's wires from less-experienced skaters, who tend to dig in their skates to keep their balance.

Thick ice also allows for a cleaner shave. When I asked about the

Oval's Zamboni, Truskolaski's face lit up. "I love my Zamboni!" He equips that sweet Zamboni with an eight-foot blade, which can resurface the whole Oval in eighteen minutes. "The 'boni can go up to nine miles per hour," he said. "But speed like that makes really poor ice."

Speed like that?

"Hey, that's fast!" he said. "But we need to go slow, take our time, and make a great sheet." He said he's still driving one of the same machines used during the 2002 Olympic Games. "If you take care of your equipment, the machine and the ice plant," he said, "they last a long time." It takes knowledge and experience to care for it all—a fact that reinforces that the ice business, whether pertaining to bagged ice or to rinks, still operates as a kind of apprentice system. There isn't a formal school that trains people like Truskolaski. He had to learn from the ice managers before him, and those experts learned from the generation before them. "At some facilities around the country, they just don't have the knowledge," he said. "They run their equipment into the ground."

The care with which Truskolaski and his colleagues treat the ice at the Oval is evident as soon as you see it. And hundreds—perhaps thousands—of people have seen it. The Oval is open to the public year-round, not only for skating but also for amateur hockey and speed skating demonstrations. "It's great to have a facility here to engage youngsters who can develop into the sport," he said. "They come in here, they take one look at the ice, and they're obsessed."

PART 4

THE FUTURE OF ICE

CHAPTER 12

Fevers, Freezers, and Frankenstein

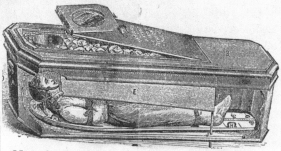

**Ad for a J. C. Taylor & Son's "Cold Air Ice Casket"
(1879)**

In May 2021, a little over a year into the COVID-19 pandemic, I sat with my spouse under a FEMA-run tent in Newark, New Jersey, waiting to receive my first of two jabs of the Pfizer-BioNTech vaccine. I knew from several recent news stories that, behind the scenes, the vaccine vials were being stored at well below freezing. That's what kept them stable and safe to be administered to patients.

Most vaccine sites stored their vials in dry ice, which had to be replenished every five days to maintain temperatures. Moreover, the containers holding the vaccines were not to be opened more than twice a day, lest too much cold escape. If medical staff failed to abide by these guidelines, the vaccines were deemed unusable and had to be destroyed.

By the end of 2021, the World Health Organization estimated that nearly half a million lives had been saved by the COVID-19 vaccine—an estimate that did not include people under the age of sixty or those indirectly saved because of a reduction in transmission. The full impact of the vaccine will take several years to determine, but in the meantime, the science makes clear that extreme cold saves lives.

Over the course of two centuries, America's obsession with ice led to a deeper understanding of cold, and that led to medical break-throughs that have the potential to save countless lives, not only from a novel coronavirus, but from cancer, organ failure, and other tragic diagnoses. This knowledge that ice—and by extension, coldness—helps the body heal has been passed down for millennia.

In 1862, an Egyptologist named Edwin Smith discovered a scroll dating to 3000 B.C. that references the use of a cold press composed of fruit, mortar, and other ingredients to heal injuries. The text doesn't mention ice or snow specifically, but that makes sense considering the locale in which it was written. Ancient Egypt wasn't exactly known for its cold. The concoction, however, likely *felt* cool and soothing on the skin. Thousands of years later, the ancient Greek physician Hippocrates theorized that cold presses applied to injured body parts could reduce swelling. By the first millennium, the philosopher and physician Avicenna wrote the *Canon of Medicine*, in which he listed snow and ice alongside opium as types of anesthetics.

Ice's influence on medicine carried into the eighteenth century, when a physician named James Currie conducted a series of controversial experiments to determine how cold affected body temperature. He immersed his human subjects repeatedly in ice water to see how their body temperatures changed. The experiment bordered on torture, but led to the discovery of afterdrop, a natural phenomenon wherein a hypothermic patient's body temperature drops again after rewarming.

Around the same time, an Austrian peasant farmer named Vincenz Priessnitz gained notoriety by healing himself with ice. While working in the fields, Priessnitz was knocked to the ground by an angry horse. The encounter broke several of his ribs. Pulling himself up on his elbows, he managed to crawl and drag himself back to the house, where he wrapped what precious ice he kept stored in his icehouse with a bandage, then applied the bandage to his bruised sides. Later, he told anyone who would listen that he was back working on the farm in just ten days, apparently having healed completely.

Today we know that a complete recovery in such a short amount of time is next to impossible, and that in Priessnitz's case, the ice likely reduced his swelling and made the pain more tolerable. But Priessnitz wouldn't have known this. What he did know is that he could make a fortune selling his ice treatment to others. He gave up farming and opened a cold-therapy spa equipped with ice pools and cold showers. His patients (he called them "guests") came from miles around and reported to feel better after their visit, and they probably did—the ice would have reduced their inflammation just as it had Priessnitz's. And who wouldn't feel better after a visit to the spa?

Priessnitz's view of cold was at odds with the European medical profession's thinking of the day. Not only was ice rarely applied to the body, but the cold was something to be avoided at all costs. Germ theory had yet to be discovered, so coldness was often assumed to be the source of illness. A cholera outbreak in 1840s England was believed to have been caused by "cold fruits" like melon and cucumber. When Europe's doctors crossed the Atlantic, they brought their views of cold with them, instilling their fear of cold into the American psyche. Just how badly did Americans fear the cold? It's telling that the American Industrial Revolution—which gave us the telephone, the light bulb, the sewing machine, and the airplane—failed to produce one machine to combat the heat.

Not until the 1860s and the onset of the Civil War did American attitudes toward cold start to evolve. The change was ushered in by the sheer magnitude of wartime death and disease. Recent estimates based on census data of the period figure that close to 752,000 Americans died during the Civil War, and countless others were left disabled. Several factors caused such a staggering number. Combat was up close and personal (a family member could quite literally be fighting another family member), and neither side had a coordinated system to get injured soldiers off the battlefield quickly. It wasn't unusual for a wounded soldier to lie in the field for days—sometimes in the rain—before someone would bring him to a hospital.

When soldiers weren't dying from their wounds, they were dying from disease. Medicine during the Civil War wasn't that different from medicine during the Middle Ages. Few physicians received formal medical training, and most of them still subscribed to the theory of the four humors, an ancient Greek system of medicine that held that four bodily liquids—blood, phlegm, yellow bile, and black bile—needed to be kept in balance for the body to be healthy. The need for sterilization wasn't yet fully understood, so camps were riddled with viruses and bacteria. Gangrene was everywhere, which was bad news for soldiers in need of surgery. Three out of every four surgeries performed during the war was an amputation, and more than 26 percent of amputees died within days of the procedure, usually because of infection.

Outside the camps, things weren't much better. In the South, yellow fever killed over ten thousand people during the war years, while the North suffered more than seventy-five thousand cases of typhoid fever. Measles killed thousands on both sides. A smallpox vaccine had been invented seventy years before the war, but most of the population was unvaccinated. Outbreaks infected both armies, with Black soldiers contracting the disease at six times the rate of whites. In wartime, vaccines in general were in short supply, so during an outbreak of smallpox, vaccine material was aspirated from the pustules of vaccinated people and given to the unvaccinated in hope of inoculating the wider population. There's little record of the success of this makeshift procedure in preventing smallpox, but it did lead to the transmission of many cases of syphilis.

With death and disease everywhere, physicians and nurses looked beyond their meager training for new therapies, anything to stop or at least slow the gush of patients that overflowed their hospitals. Among them was the use of ice.

Generals of both armies considered ice a necessity for their soldiers. It was something to soothe the tongue on a hot day. In the first year of the war, army hospitals north of Washington, D.C., were allotted one half pound of ice per patient, while those to the south of the city were allotted one pound each.

Some doctors used it to reduce swelling or fevers, but many still regarded ice with suspicion. One Union Army doctor observed several soldiers drinking cold water after a march through extreme heat, only to die mere hours afterward. The loss of life, he wrote, "was greater than that general had previously experienced in any battle." He blamed their deaths on the cold water. Another doctor watched as a group of soldiers, having just marched through the summer heat, stopped to enjoy "copious use of cold drinks." Within hours, he wrote, the men were "seized with dimness of sight" and "vertigo" before falling to the ground, their faces "suffused with blood." He, too, blamed the ice that had floated in their cups.

A few open-minded doctors saw ice as a means for providing comfort, and an even fewer number touted the stuff as if it were a miracle drug. One such doctor described a patient suffering from diphtheria, which swells the neck glands and causes a thick, gray film to form over the back of the throat. Having tried everything else, the desperate doctor applied ice compresses around the neck of his patient for a week. By the eighth day, the patient had recovered fully, probably due to sheer luck, though the ice likely made him more comfortable. The doctor, however, became a true believer in the healing powers of ice. "Should the ice prove as beneficial in other cases as in this," he wrote, "it would be a great blessing for mankind."

The fervor with which that small subset of doctors touted the benefits of ice spread throughout both armies, and by the war's second year, ice had become a precious commodity on both sides. The Union Army set up blockades to prevent ice merchants from supplying southern cities. And military strategies evolved to target and attack the enemy's ice reserves.

One of the best-documented attacks took place in January 1864. It was a colder-than-usual winter, with temperatures dangerously below zero. The Arkansas River had frozen to depths rarely ever seen, and the Union Army, which had only recently pushed its way into Little Rock, decided to harvest it. It would provide more ice than those men had ever seen in one place. They stored the ice in a commandeered icehouse at a nearby hospital, to which Confederate patients were likely denied access. When temperatures rose somewhat a week later, the Union Army loaded the ice onto a train for shipping north, where it would be welcomed by and distributed among the soldiers, many of them now in their third year of fighting. But just as the train started chugging, the sound of hundreds of horses galloping and men shouting carried over the hills. A hidden Confederate unit bolted toward the train and, wielding every kind of metal tool they had found, from hoes to pitchforks, tore up the tracks, rendering the train immobile. The ice, lamented a Union doctor who had watched the whole thing from the hospital, "melted in the cars."

When the war ended, the medical community continued to experiment with ice, and by the turn of the twentieth century, doctors were asking not how to avoid the cold but how they could make it work in the patient's favor. A medical student named Temple Fay was especially interested in how cold affected the human body. He asked a professor why cancers were rarely found below the knees and elbows, but the professor didn't know—no one did.

Fay conducted a series of experiments that revealed that lower temperatures reduced the growth of cancerous tumors, and since the ends of limbs are cooler than the body's core, cancer didn't grow there as frequently as on the trunk or head. Fay graduated medical school and in the decades that followed conducted some of the first experiments to be officially classified as therapeutic hypothermia, the process of deliberately lowering a patient's body temperature for the purpose of treating disease. Between 1938 and 1940, Fay reduced the body temperatures of 126 patients diagnosed with cancer to a mind-boggling seventy-five degrees Fahrenheit by soaking them for hours in ice. Most lost consciousness during the soak, but all of them survived.

The first of these patients was a young woman with metastatic breast cancer, who was living with debilitating pain. Fay started his treatment by repeatedly applying a cold press to one of the breast tumors for several weeks. He did this until biopsies showed that the tumor was getting smaller. Buoyed by these results, he decided to reduce her entire body temperature with the hope that doing so would reduce the size of her other tumors. That winter, he put his patient in a tub of 150 pounds of cracked ice, turned the heat off in the room, and opened the window so that the room filled with freezing air. She sat there all morning, her temperature dropping. By late afternoon, her core body temperature was in the low nineties. Her respiration had slowed and her pulse was barely detectable. After eighteen hours, Fay rewarmed his patient, and she fully regained consciousness, her pain greatly reduced. More than likely, the extreme cold had reduced inflammation and created a numbing effect that made her more comfortable.

Fay went on to "ice" several more patients, nearly 10 percent of whom died from the extreme cold—a number he felt comfortable with because, he told them, their chance of survival was eight to one. Of those that survived, 95.7 percent reported a significant reduction in pain.

Despite such promising results, Fay's work was brought to an end by a rebellion among nursing staff, who believed that his "cold ward" was too dangerous for patients and too demanding on the nurses. Unable to convince them otherwise, he focused on a different kind of ice therapy—cryosurgery, a mildly invasive procedure that involves freezing human tissue with ice crystals. If cold had such a positive effect when applied to the outside of bodies, what miracles could it work if it was injected? He developed tiny metal capsules containing a refrigerant and implanted them into the brain tissue of consenting patients with brain lesions. The treatment showed promising results. He noted that in areas surrounding the ice-cold capsules, inflammation and infection were absent. These discoveries led Fay to develop the first deliberate program of hypothermia for traumatic brain injury, a treatment that is still being studied—and to an extent used—today.

Fay's experiments would have continued, but they fell into the wrong hands. In 1940, German spies stole a copy of Fay's manuscript describing his experience with hypothermia. They used it to devise their own experiments, searching for ways to ensure their airmen would survive when shot down and stranded in the freezing North Sea. Nazi scientists conducted these experiments on approximately three hundred concentration camp prisoners, who were put under heavy sedation and then placed in tubs filled with ice water for several hours. Some of the victims' heads were submerged, causing damage to the brain stem. Almost all the prisoners subjected to the experiments died.

When these horrors were revealed to the world, doctors scaled back their hypothermia research, and Fay discontinued his research altogether. In 1944, he wrote:

> The wide application of cold therapy almost 100 years ago, when ice was a luxury, reflects today that ever human tendency to ignore what is plentiful, common, and easily at hand. The field of refrigeration or hypothermy is broad and deep, awaiting exploration by those who have modern facilities.

By the 1960s, the stigma of the Nazi experiments began to wear off, and research resumed. Neurosurgeons in particular were interested in Fay's work; they hoped it could help treat inoperable aneurysms and brain lesions. The most infamous of doctors to pursue this field of study was Dr. Robert White, who established the Brain Research Laboratory (BRL) in 1961 at Case Western Reserve University in Cleveland, Ohio. White, who graduated from Harvard Medical School, was mostly interested in whether therapeutic hypothermia could prevent serious brain damage after head trauma.

In the decades between Fay's and White's research, the medical field changed significantly. Experiments that had once killed up to 10 percent of participants were no longer legal. Unable to experiment on humans, White turned to animals, quickly becoming a target of animal rights activists. To be fair, his experiments were really, really weird. Horrific, even. They also led to some of the most significant breakthroughs in neuroscience.

White used hypothermia to lower the body temperatures of monkeys, then surgically removed their entire brains. He kept the brains "alive" with artificial blood flow. In a documentary film of this procedure, a disembodied monkey brain can be seen pulsing with blood circulating through a series of connected metal tubes. These monkeys, existing only as brains, remained "alive" for up to three hours. Over the next decade, White extended the brains' existence up to forty-eight hours.

The question of whether these brains were in fact "alive" was the purview of philosophers, perhaps the clergy. White wasn't interested in answering it. He wanted to know instead whether the brains maintained any level of awareness. His experiments were inconclusive. Then, in 1970, White expanded the experiment again, this time using whole heads.

With a team of thirty doctors, nurses, and aides, he lowered the body temperatures of two monkeys, and then surgically removed both of their heads. He connected the recipient monkey's head to the circulatory system of the donor monkey's body. When he rewarmed the head and body, an EEG showed, miraculously, horrifically, neural activity. The monkey opened its eyes and looked around. It could also bite and swallow, but because its head lacked connection to a spinal cord, the animal was paralyzed from the neck downward. The monkey lived for thirty-six hours. In a 2007 interview, White proclaimed that the test subject "was not a happy monkey."

In the BRL's final days (it closed in 1996), White was using cerebral hypothermia to supercool the brain to minus forty degrees Celsius using specially designed ice crystals. His work was motivated, he said, by patients with debilitating spinal cord illnesses, and he was recognized for his contributions to medicine by twice being nominated for a Nobel Prize. The nominations did little to improve his reputation among the public, but he didn't seem to care. In fact, White seemed to welcome controversy. In a 1999 article for *Scientific American*, he wrote, "I predict that what has always been the stuff of science fiction—the Frankenstein legend, in which an entire human being is constructed by sewing various body parts together—will become a clinical reality early in the 21st century." Picking up on the comparison, a lawyer later referred to White as "Dr. Frankenstein." He likely meant it as an insult, but the comparison was, literarily speaking, rather apt. At the end of Mary Shelley's novel, Dr. Franken- stein's monster flees to the Arctic Circle, where he laments, "I was better fitted by my conformation for the endurance of cold than heat."

Today, the practice of therapeutic hypothermia is establishing itself in less controversial ways: to treat intracranial swelling and spinal cord injuries and, most often, to protect the brain just after cardiac arrest. Still, the practice is not common and is considered ex- perimental. There is not enough research to support standardized cooling procedures. In the United States, therapeutic hypothermia is practiced in only 17 percent of hospitals nationwide.

Despite its rarity, therapeutic hypothermia has proved itself to be an invaluable clinical tool. One of its most well-known practitioners is Dr. Joseph Varon. Originally from Mexico, the cardiovascular spe- cialist practices in Houston, Texas, and is a member of the Royal So- ciety of Medicine. Back in 2005, a fifty-four-year-old drowning victim was flown to Varon's emergency room. The patient had been swimming in Mexico when a giant wave pummeled him under, suffocating him and crushing his neck. "He was almost dead to the point that I was about to declare him brain dead," said Varon in an interview at the time. But something told Varon to hold off. Instead, he took a rather unorthodox approach. He "froze" the patient by applying a thera- peutic blanket that lowered his body temperature to ninety degrees

Fahrenheit. His hope was that the cold would preserve the patient's brain and organ tissues, giving the rest of his body time to stabilize. After three days, the doctor warmed the patient up and was relieved when he was given a sign that his plan had worked: the patient smiled.

In 2013, Varon once again had the opportunity to employ therapeutic hypothermia, this time to a stroke victim. The patient was himself. Varon knew the signs of a stroke and how such an event can devastate the brain and body. Having seen firsthand the power of cold to preserve tissue, he wrapped himself in cooling blankets and took himself to the emergency room. He spent weeks in the hospital but survived with his brain function fully intact. At the time of this writing, Varon was still practicing medicine, serving as the chief of critical care services at the United Memorial Medical Center in Houston.

Therapeutic hypothermia is just one legacy of the use of ice in medicine. Cryosurgery—also called cryotherapy, cryogenic surgery, and cryoablation—is another. It's defined as the "controlled destruction of tissue by freezing" and is used in a range of hospitals and office-based practices to treat all kinds of ailments. With its long, sharp needle, a cryosurgery procedure looks like something out of a horror movie, but it's a relatively painless process. In dermatology, cryosurgery is used to treat warts and other benign lesions because the resulting wound heals fast and leaves little scarring. It's also used in the treatment of some skin cancers with impressively high cure rates. In ophthalmologists' offices, cryosurgery is used to treat retinal detachment and some eyelash disorders. In gynecology, the method has been used to treat cervical cancers, and it is gaining popularity with surgeons who treat prostate, rectal, and kidney cancers.

To treat cancer, the process begins with the insertion of a needle called a cryoprobe through the skin and into the cancerous tumor. A gas—usually nitrogen or argon—is pumped into the needle, cooling it to at least seventy degrees below zero. When the needle enters the tumor, it forms ice crystals inside the cancerous tissue. Those crystals are allowed to thaw, and then the process begins again, freeze, thaw, repeat, until the cancerous tissue starts to die.

As with any invasive medical procedure, there are documented risks associated with cryosurgery, but it is generally less intrusive than surgeries involving knives, and it rarely harms the healthy tissue surrounding the tumors. This is probably good news for any patient who's experienced severe scarring after the removal of a tumor. And yet when I reviewed the Mayo Clinic's list of cancers most often approved to receive this treatment, I saw only bone, cervical, eye, kidney, liver, lung, and prostate. Noticeably absent was breast cancer, which surprised me. Aren't breasts easier to, well, puncture with needles than are lungs or livers?

I emailed an interview request to the National Cancer Institute, hoping to learn more, but its response was short and to the point: "As you may be aware, cryotherapy is not a standard treatment for breast cancer, but is being investigated in clinical trials."

Unsatisfied, I researched further and discovered the work of Bailey Surtees, a graduate of Johns Hopkins's biomedical engineering program. Along with fellow Hopkins graduate Clarisse Hu and Nicholas Durr, a Johns Hopkins assistant professor of biomedical engineering, she cofounded Kubanda Cryotherapy, a biomedical upstart that seeks to bring cryosurgery to underserved populations to specifically treat breast cancer. Surtees said she'd be happy to talk.

Surtees, who's originally from Oklahoma, grew up deaf in her left ear. She said that a cochlear implant wouldn't have helped her condition, but learning about the device from doctors sparked a lifelong fascination with how engineering can support the human body. Getting into the Johns Hopkins program was a "dream come true," she told me. After moving to Baltimore to attend the school, she stayed in the region, working with professors and other graduates of the program to "put good into the world" through cryosurgery research.

That research led to Kubanda, which has designed and produced working models of a portable cryosurgery machine that she describes as "rugged and purely mechanical," because "when you hook up electronics, everything requires a lot more maintenance." The product is purposely designed and based, she said, on "high-school-level chemistry." The control system hooks on to a gas tank, which is connected to a handheld unit via a hose. The gas tank sends room-temperature carbon dioxide through the hose at an extremely high pressure, which cools it considerably. When the gas reaches the probe attached to the handheld part of the machine, the needle drops to minus seventy degrees Fahrenheit. The machine's simplicity and portability mean that it could, theoretically, be used almost anywhere.

"When we started our company," Surtees told me, "we knew we wanted to focus on underserved populations, especially lower- and middle-income countries. Here in the U.S. we have a ninety percent and above five-year survival rate [among people diagnosed with breast cancer], but in a lot of the world it's as low as twenty or even thirteen percent. We know that [the cancer] can be treated, but the treatment isn't accessible."

This machine, she said, could make cryosurgery possible in these places. "We're still five to ten years out," she said. It takes that much time "to go through the different regulatory and clinical trials to get to actual human patients." In the meantime, the machine is being used at veterinarian hospitals to treat pets with cancer. "Animals are another underserved population," she told me. Often animals are used in medical research, but there aren't "a lot of treatments that come back to the animal sector" to actually treat the animals.

I told Surtees about how little my research had turned up regarding the use of cryosurgery for treating breast cancer in the United States, and she didn't seem surprised. When it comes to breast cancer, the therapy "has stayed at the cutting edge of medicine," and mostly for reasons that have little to do with the treatment's effectiveness, she said.

"Since cryotherapy was developed specifically for high-income markets [like wealthy areas in the United States]," the system, regardless of what kind of cancer it's treating, "has to be compatible with MRI so that it can be used in advanced surgical centers." That makes cryosurgical procedures for cancer extremely expensive. "Cryotherapy systems as they currently exist are upwards of a quarter million dollars," she said, "and each cryoprobe can cost up to two thousand dollars, and you use it once."

When I asked her if her company's more portable, less expensive machine will make the treatment more available in the States as well as in other world markets, she shrugged. "Time is super important to procedures. And one of the reasons why [our product] is not as compatible in the human market in the U.S. is that our procedure takes a few minutes longer than the [nation's] state-of-the-art system." She shook her head. "Operating room time is upwards of fifty dollars per minute in America, and there's not a lot of financial pressure to bring [those] costs down." Another obstacle to implementing Surtees's machine in American hospitals is the fact that the machine was purposely designed to operate without the use of MRI, so that it can be implemented in less than state-of-the-art treatment centers. But what about ultrasound, I asked, which tends to be less expensive and more widely available, especially in rural areas?

"We've talked to hospital systems that don't have [cryosurgery]," she said, and one of the reasons they've given is that "if they switched from [more traditional forms of] surgery to cryo, then the cryo would be done by the ultrasound specialist instead of the surgeon, and the surgeons believe they should be the ones doing the treatments. So there's politics involved." Of course, she said, "we don't even need to use ultrasound, because the breast is really easy to palpate and see where the ice has grown just by touch." Still, few hospital systems in the United States will adopt an invasive treatment, even if mildly invasive, without some kind of imaging technology to guide the probe.

An even bigger obstacle is health insurance. Since cryosurgery hasn't caught on in the United States, "insurance hasn't coded for it well," Surtees told me. It's a chicken-and-egg problem: without the widespread availability of cryosurgery, insurance won't cover it, but if insurance won't cover it, the treatment isn't likely to catch on outside of a few high-income areas.

My interview with Surtees left me feeling simultaneously hopeful and despairing. On the one hand, ice, which has proved to be beneficial to the human body for thousands of years, can now be used to treat cancer, one of the deadliest diseases humanity has ever faced. On the other hand, just as harvested natural ice was once a luxury for the wealthy, cryotherapy remains accessible to only a few Americans—and to even fewer people around the world. Surtees isn't the only expert pushing for cryotherapy to become mainstream, however. "After nearly two centuries, the technique of cryotherapy remains widely applicable," wrote Dr. S. M. Cooper and R. P. R. Dawber in the

Journal of the Royal Society of Medicine. "At a time when surgical excision is in the ascendant this simple method, with its cosmetic and functional benefits, should not be neglected."

Change, however, is happening at a glacial pace. Cooper and Dawber published those words in 2001.

CHAPTER 13

TV Dinners, Henry David Thoreau, and a Swiftly Warming Planet

Mrs. Pleas Rodden puts fresh milk in her Frigidaire in
West Carroll Parish, Louisiana (1940).

On an unusually cold night in March 1845, twenty-eight-year-old Henry David Thoreau was writing in his cabin by Walden Pond near Concord, Massachusetts, when he heard a group of men shouting and what sounded like metal scraping across stone. He threw on a coat over his nightclothes and stormed outside, an icy wind slamming the door shut behind him.

The moon was full and reflected off the frozen lake, illuminating the source of the commotion: two dozen men were sawing blocks of ice out of the lake, using horse-pulled metal plows to haul the blocks onto wagons. The men were employed by Frederic Tudor to harvest the late-season ice. "Near the end of March," Thoreau wrote in *Walden*, "the ice in the pond was not yet dissolved."

Ice harvesters amused Thoreau, inspired him even, but they downright annoyed the writer's close friend Ralph Waldo Emerson, who carped in a letter that these "Irishmen" would take "10,000 tons of ice from the Pond," thus draining Walden and lowering his property value. Knowing that this was physically impossible, Thoreau looked at the situation a different way. He remembered his many baths in the lake, and wondered how his body might taste to "the sweltering inhabitants of Charleston and New Orleans, of Madras and Bombay and Calcutta" who in a matter of weeks would be buying Walden ice.

Today, the long, cold winters that kept Tudor in business are no more, and Walden Pond rarely freezes at all, let alone to a thickness that could support the weight of a team of ice harvesters and their horses. As Richard B. Primack writes in *Walden Warming*, "Climate change has come to Walden Pond."

It's come for all of us. The latest report by the Intergovernmental Panel on Climate Change warns that global greenhouse emissions must peak by 2025 and be nearly halved in the same decade if humans are to keep planetary warming below 1.5 degrees Celsius, the temperature at which climate scientists say Earth will reach a climatic tipping point.

Most appliances in American homes contribute to the CO_2 problem, but refrigerators are among the worst culprits. Refrigerators and stand-alone freezers are responsible for approximately 10 percent of an average American household's total energy use. If the fridge includes an ice maker, energy usage goes up by another 10 to 20 percent, because those ice makers never shut off—that's why they can spew out ice anytime, day or night. And that's just households. Refrigeration is responsible for 40 to 50 percent of a typical grocery store's total energy usage. In restaurants, it draws up to 15 percent. A look at energy use around the globe reveals that the cooling industry (refrigerators, freezers, and air conditioners) accounts for almost 10 percent of all CO_2 emissions. That's about half the amount produced by the airline and shipping industries combined.

How did refrigerators become such an energy suck?

During the 1920s, electric refrigeration was still a novel concept. But by 1951, they could be found in 80 percent of American households. This was one of the fastest lifestyle transitions in the nation's history. By comparison, only 9 percent of the country owned TV sets that year, though the technology had been around since the 1920s. (The number of TV-owning families grew to over 85 percent by the end of the decade, however, and their love for frozen TV dinners grew, too. By 1960, owning a fridge and a TV came to signal that one had arrived to the American middle class.)

Throughout the history of the electric fridge's development, several actors—including the ice industry, the power industry, and state and federal governments—played roles in shaping how refrigerators function and draw energy. One hundred years ago, global warming wasn't on the minds of these industry insiders, because the science wasn't widely known and the concept had yet to pierce public consciousness. On their minds instead was profit, as well as progress. The ice industry had already shown Americans the benefits of keeping their perishables cold. Now, the refrigeration industry would show them just how much better electric cold could be.

––––––––––

In 1929, the Roaring Twenties descended into the Great Depression, and America's economy cratered. That decade also brought about a new era of refrigeration. By the mid-1930s there were officially as many electric refrigerators in use as iceboxes in the country, and technology in that sector was improving fast. Just forty years earlier, the electric refrigerator had been out of reach to most Americans because it was so expensive to manufacture. In 1885, *Scientific American* editors wrote, "We know of no cheap way of making a refrigerator for household purposes without ice. Such a process, if convenient, would be worth a great deal of money." They were more right than they probably knew.

Commercial refrigerators were also in use by then, but they were large beasts of machines, weighing as much as 5 tons each and requiring around-the-clock maintenance. Engineers found refrigerators difficult to shrink because of efficiency problems: the smaller they were, the more heat they generated, counteracting the cold they were purchased to create. Not that inventors didn't try to make them work. "The small refrigerator plant, or ice machine, is coming," wrote Mary Pattison in her 1915 work *Principles of Domestic Engineering.* "We already have a number from which to choose, but up to date we can not fit them in practically to the average home with assurance of economy."

The first electric fridge-like device hit stores in 1914. The DOMELRE (short for "domestic electric refrigerator") was built to be installed in existing iceboxes. Once plugged in, it would cool the air inside the icebox using sulfur dioxide instead of ice. The unit came with ice-cube trays, the first on the market, which were the machine's biggest draw: it made having ice possible without the cost of delivery. Between 1914 and 1922, stores sold a few thousand of these little machines, but not nearly as many as they did iceboxes.

Two years after the DOMELRE's debut, the Packard Motor Car Company introduced a device that used similar technology, but it sold hardly better. In the years that followed, General Motors and other large companies attempted to enter the electric refrigeration market with similar apparatuses, but none had any staying power. They broke often, and because the shapes and sizes of iceboxes were never standardized, the little machines didn't fit very well. Stand-alone refrigeration units followed, but still, sales remained stagnant. Not until the mid-1930s did manufacturers come to understand why their products weren't flying off shelves. They faced a problem similar to what Tudor had faced when he first attempted to convince people to buy ice: no one knew how such a product would fit into their lives. The public needed to be sold the *idea* of refrigeration before they could be sold the invention.

Thus followed an advertising campaign so expansive it rivaled the size and reach of those launched by the automobile industry, the other lifestyle disruptor of the era. Ads for electric refrigerators sold more than their ability to chill—they sold a sense of awe. They depicted refrigerators as a miracle of a machine that didn't require ice to keep cool. In fact—can you believe it?—it could create cold on its own. A 1920s ad in the *New Yorker* read, "A little water is put in some mysterious place, a few minutes pass, a magic door opens, and a tray of small ice cubes appears before your startled eyes." Other ads depicted families standing around their new appliance, mouths agape at the marvel before them. In *Refrigeration Nation*, Jonathan Rees writes that the recurring image suggests "the Magi and the Christ child."

As much as they inspired wonder, refrigerators were also a source of great annoyance because they frequently broke. Their engineering was poor, and manufacturers had yet to standardize their appliances, making a broken fridge expensive and difficult to fix. A memo in the Frigidaire archives describes an instance when a company repairman visited the wife of a General Motors executive to fix a sulfur dioxide leak coming from her fridge. The volume of gas quickly overwhelmed him, and as he stumbled his way out of the kitchen, he tripped and startled the gardener, who attracted the attention of the whole neighborhood with his cries for help. This is why, concludes the memo, the company's "servicemen acquired strange reputations."

———————

The fledgling industry gained its wings when General Electric entered the market in the late 1920s, bringing mass-produced, well-designed, and mechanically standardized appliances to consumers. For GE, however, fridges weren't the end game—they were a means to something bigger. The fridge of the 1920s was notoriously energy inefficient. Like today's refrigerator, it was always powered on and drew more energy than any other household appliance. According to one 1926 study, by simply installing an electric refrigerator, an average family doubled the amount it owed the electric company each year. The more fridges that GE sold, the greater the boost to its bottom line.

The GE fridge was expensive to operate, yes, but its energy costs were at least predictable—a feature that the ice industry couldn't compete with. By comparison, the costs and supply of ice were erratic and changed with the seasons. With a new electric refrigerator, all a family had to do to get ice was open the door to their newfangled appliance. Such uninterrupted access to ice was so new, so exciting, that many early fridges were sold with recipe books that included novel ways to prepare ice, including freezing cherries and mint into cubes for special occasions.

As the nation sunk deeper into the Depression, government interventions helped fuel America's growing fascination with electric refrigeration. The National Housing Act of 1934 allowed households to take out government loans to purchase fridges. The Rural Electrification Act of 1936 brought electricity to the most remote parts of the nation, further enticing Americans to purchase electric refrigerators. By 1947, only 30 percent of American households still had iceboxes. Just six year later, in 1953, the last icebox manufacturer in the United States shuttered its doors.

No other country took to refrigerators quite like America, and that love grew even deeper with the introduction of the electric freezer. Large, frozen storage units held at shared public facilities arrived first. They operated similarly to how safe deposit boxes work in banks. Hungry families went to the freezer warehouse, turned a key in their designated locker, and either put a perishable in or took one out. In 1941, the cost to rent a locker was approximately ten dollars per month, a price too steep for most American families. Then came World War II, which brought food shortages that forced families to stock up on perishables, and rentable freezers became something close to a necessity. The National Frozen Food Locker Association estimated that, by the 1940s, families on average were making trips to their frozen lockers at least once per week.

In the early '50s, freezers grew smaller, and the industry's marketing teams designed ads for domestic appliances that appeared in all the popular magazines and newspapers of the day. A Ben-Hur Freezer ad from 1952 crowed about its "roomy interior," which had "plenty of room for hundreds of pounds of fresh frozen foods." The appliance also made guest appearances in popular entertainment. In a 1952 episode of *I Love Lucy* titled "The Freezer," Lucy dreams of having a freezer of her own. When Ricky denies her request, neighbor Ethel saves the day by snagging a commercial-sized freezer for free from her butcher uncle. Having lost their heads with so much excitement, the women order way too much meat, and as Lucy works to stuff it all in, the freezer door slams shut, locking her inside. It's the perfect setup for Lucille Ball, allowing her to show off her physical comedy chops. But the most notable aspect of the episode is the husbands' palpable excitement at the thought of seeing a freezer in their apartment building for the very first time.

The industry sold one million freezers that year. The year following, Americans owned over three million. This growing appetite for the appliance led to another innovation, the prepackaged frozen meal, or as the Swanson food company called them, "TV dinners."

Within twenty years, freezers and TV dinners alike became so ubiquitous that jokes about their tastelessness no longer needed elaborate setups. In a 1973 episode of *Maude*, the titular star serves her husband, Walter, a Swanson frozen dinner just after an argument. When he complains that his meal isn't homemade, Maude quips, "Walter, if you think *that's* a frozen chicken, wait till you see what you find in bed tonight." By the 2010s, Jerry Seinfeld managed to eke out a two-minute stand-up routine on the quality of Swanson's Hungry-Man frozen dinners. A "little taste of prison right there in your home," he joked.

———————

In the 2020s, at least one in four Americans owned at least two refrigerators. Freezer aisles are in almost every American gas station and grocery store, and even the cheapest motels provide small refrigerators in every room. Like the ice trade before it, the electric refrigeration industry fundamentally changed how Americans eat and store food. It also increased the nation's energy needs. Between the end of World War II and 1975, the amount of electricity consumed by refrigerators grew by more than 350 percent. At the time of this writing, most American households still rely at least partially on fossil fuels for energy, and contemporary refrigerators, while technically more energy efficient than their early twentieth-century predecessors, still draw an inordinate amount of power.

And it's not just the amount of power they draw that contributes to global warming. Contemporary fridges and freezers use chemicals, not ice, to create their cooling effects—chemicals that damage the planet. Throughout the first two-thirds of the twentieth century, the chemicals most often used in fridges were chlorofluorocarbons, more widely known as CFCs, which scientists discovered were depleting Earth's ozone layer. World leadership listened to the scientists' warnings, and at the 1987 Montreal Protocol, two hundred countries signed an agreement to phase out CFCs and shift to hydrofluorocarbons (HFCs), which break down the ozone less but are still extremely potent greenhouse gases. These chemicals are still used today.

Fridges become even less environmentally friendly as they age and break down. According to Project Drawdown, a nonprofit that analyzes the effectiveness of proposed solutions to climate change, almost 90 percent of refrigerant emissions occur at the end of an appliance's life. This problem is compounded by the fact that there are few laws dictating how old fridges should be disposed of. While many states employ services approved by the Environmental Protection Agency (EPA) or utilize local authorities to pick up and recycle old fridges, there's nothing on the books at the local level that says a fridge owner can't simply take the old appliance out back and strike it with a hammer, smashing it to bits and lettings the chemicals seep into the ground.

These have been known problems for some time. During the 1970s energy crisis, California became the first state to regulate refrigerator efficiency. Those regulations were so influential, models that didn't conform to the state's standards soon disappeared from the market, and in 1978, Congress voted that national standards were necessary and passed the National Energy Conservation Policy Act. It directed the Department of Energy (DOE) to develop standards for refrigerators and twelve other kinds of appliances. By 1980, it had a set of proposed standards even more stringent than California's. However, appliance manufacturers with a strong lobby in Washington strongly opposed national standards, so they were never implemented. By 1982, the Ronald Reagan administration was in power, and the once pro-standards DOE reversed course. The department called for the elimination of standards altogether, arguing that the free market would bring about improvements in efficiency equivalent to standards mandated by government.

Congress, however, continued to push for standards, and in 1986, the House and Senate both passed the National Appliance Energy Conservation Act. Reagan, however, refused to sign it. In his memorandum of disapproval, the president cited the impediment of the free market as the bill's biggest fault, that it "limits the freedom of choice available to consumers who would be denied the opportunity to purchase lower-cost appliances." Reagan's administration was a loud proponent of free-market theory—in all aspects of governing but especially in energy consumption—but there was little empirical evidence to support that stance.

In fact, when it came to increasing energy efficiency, studies showed exactly the opposite. In his groundbreaking 1983 study on the matter, David B. Goldstein, codirector of the Natural Resources Defense Council's Climate and Clean Energy program, revealed that energy efficiency "ranks last among various features" consumers care about most, and that when presented with the option to pay more upfront for an energy efficient refrigerator in order to save more in energy costs in the long run, consumers didn't bite. They preferred to spend less in the moment, because long-term savings felt theoretical. Government-supported standards, meanwhile, had already proved in California that they saved consumers a huge amount of money.

Goldstein argued further that standards would save fridge manufacturers money as well. "Manufacturers today face a great risk in designing [a future] product line," he wrote. "How much emphasis should they place on efficiency? A wrong decision—to produce efficiencies that are either too high or too low—will be costly. Efficiency standards, by contrast, give manufacturers a known target, based on engineering requirements rather than uncertain sociological phenomena."

In 1987, just one year after refusing to sign the National Appliance Energy Conservation Act, Reagan agreed to sign a watered-down version. By this time, states other than California had already begun adopting their own standards, because in addition to helping the environment, energy standards forced other kinds of standardization that eliminated engineering malfunctions. And fewer broken fridges meant more and happier consumers.

Over the next few years, as environmental news grew more despairing, energy standards grew stricter. In 1992, the EPA introduced Energy Star as a voluntary labeling program designed to help consumers identify the most energy efficient appliances, including refrigerators. The problem, however, was that Energy Star products were self-certified by third parties, leading to inaccurate labeling. It took almost twenty years, but in 2011, the EPA took over the certification process, and the industry as a whole made huge strides in energy efficiency. In 2021, Energy Star updated its ratings system. The new system was stricter than the old one and forced a reclassification among appliances that resulted in some being labeled less efficient than they had been previously. The change will, in theory, put pressure on manufacturers to keep improving the efficiency of their appliances. But at the time of this writing, a significant problem still remains with energy labeling: the process of categorization isn't nuanced. Types of fridges are grouped together and given energy ratings without consideration of how tiny differences in parts and engineering can cause their energy use to vary widely from one manufacturer to another. It's possible to buy a refrigerator with a top Energy Star rating that's actually, as a reporter from *Inside Climate News* put it, a "carbon bomb."

Wanting to learn more, I reached out to Avipsa Mahapatra, a climate campaign lead at the Environmental Investigation Agency. The problem with energy labels, she said, is that "there are so many constraints and physical elements that determine how efficient a product is. What you think you might be getting is not [necessarily] what you might be getting. The whole energy efficiency standards sector needs to be more detailed." She continued, "But the fundamental challenge here is to introduce simplicity by using complexity. There's a constant tension between making sure that you are catching every different element while keeping it simple enough for consumers." In other words, make the system too complicated, and too few people will even bother trying to navigate it.

Another obstacle to energy efficiency is politicization of the issue, which grew worse during the Donald Trump administration. In 2020, at a rally held at the University of Wisconsin–Milwaukee Panther Arena, the president evoked nostalgia for household appliances of yesteryear by comparing them to those we use today. "Sinks, toilets, and showers—you don't get any water," he said. He then railed into dishwashers and refrigerators. The rally took place on the heels of his administration's rollback of several energy-efficiency standards and the establishment of new ones by executive order. Some of these new standards were so low that even manufacturers didn't want them.

So what hope is there for improving refrigerator energy standards? I posed that question to Ralph Cavanagh, the energy codirector of NRDC's Climate and Clean Energy program. Cavanagh has been studying and writing about refrigeration energy standards since the 1980s. His response was surprisingly optimistic. "Today, refrigerators use less than one-fourth as much electricity as they [once] did," he told me. "They're much better machines. They deliver far better service and they cost less." This was achieved, he explains, through government standards, but also through something called the "golden carrot program." In 1991, a national coalition of utilities pledged to award a refrigerator manufacturer $30 million if it could develop a fridge that would exceed 1993 federal energy standards by at least 25 percent. The award was given to Whirlpool Corporation, which brought some of the world's most energy efficient fridges to the market. The increase in energy use we're seeing from refrigerators, Cavanagh said, was because more people than ever before are using them—not because the fridges themselves are getting less energy efficient.

Change is also happening around how manufacturers are thinking of environmentally damaging chemicals. While the switch from CFCs to HFCs was a start, it wasn't good enough, argues Mahapatra. "We were ill-advised to put a Band-Aid on the problem," she said. "To change from one class of chemicals to another one was only slightly less bad for our planet." Mahapatra's work—and the work of other activists—is finally paying off, because once again, world leadership is starting to recognize the problem. In 2016, officials from more than 150 countries signed the Kigali Amendment, an agreement to reduce HFC consumption by 80 percent by 2047. The United States Senate signed in 2022. Given that CFCs and HFCs have accounted for close to 11 percent of total warming emissions to date, this amendment, if achieved, could avoid more than 0.4 degrees Celsius of global warming by the end of the twenty-first century.

There's also a growing movement to replace these chemicals altogether with something called natural refrigerants. This class of chemicals includes ammonia and propane, both of which have comparatively low global warming potential. Some major brands such as Coca-Cola and PepsiCo are already using natural refrigerants in their commercial refrigerators. When asked why more companies and manufacturers haven't made the switch, Mahapatra cited safety concerns: ammonia is highly toxic and could present a health risk if it escapes through a leak, and propane is flammable. But those concerns, she continued, were largely conjured up by fridge manufacturers to avoid having to make the switch. "Unbeknownst to a lot of people," said Mahapatra, "[safety standards] are actually written by the industry."

Natural refrigerants are just one kind of cooling alternative, however. New technologies have the potential to radically change how fridges operate. Today's typical fridge works using a process similar to what John Gorrie used in his ice machine prototype back in the mid-nineteenth century. A compressor puts pressure on a gas, raising its temperature and pushing it into the coils located at the back of the appliance. In the relatively cooler air of the kitchen, the hot gas in those coils releases some of its heat and becomes a liquid. As that liquid flows back into the sides of the fridge, the pressure is removed and the liquid evaporates into a gas, absorbing the heat out of the fridge and cooling it in the process. Some scientists are experimenting with doing away with liquid coolants altogether.

One alternative is a fridge that utilizes something called magnetocaloric material, a substance that absorbs or emits heat when exposed to magnets. Through a series of magnetic fields, the material is repeatedly expanded and compressed to absorb and release heat. This magnetocaloric effect was first discovered by Pierre Weiss in 1917, and after more than one hundred years of research, magnetic fridges are finally being manufactured. These fridges are more costly than those with more conventional mechanics, but as demand for better energy efficiency grows—and according to Cavanagh, it is growing—so, possibly, will demand for these kinds of fridges. Haier introduced a magnetic wine cooler in 2015, and GE and Whirlpool are working on similar technology. A recent industry report states that the magnetic refrigeration market is expected to double in the first half of the twenty-first century.

Another potentially game-changing technology utilizes material called plastic crystals. Excited by this research, I called Dr. Xavier Moya, an assistant professor at the University of Cambridge who has published work in this area. Plastic crystals, he told me, are a solid material, but they aren't actually made of plastic. "Though they are found in plastics," he said, as well as in paints and some cosmetics. The crystals get their name from the plasticity of their molecules, which, when subjected to extreme pressure, move toward one another. This movement draws heat from their surroundings, a process that, if it were applied to refrigerator technology, would mimic how liquid coolants work. "Because [plastic crystals] are solids," he told me, they are "not going to leak and go into the atmosphere." This makes them a much more environmentally friendly alternative to liquid chemicals. "The challenge that we have at the moment," he said, "is building a prototype that demonstrates high energy efficiency as well. If we can do this, then that's potentially a transformative technology." Moya is currently working with Cambridge Enterprise, the commercialization arm of the University of Cambridge, to produce a working prototype.

Work like Moya's might very well transform the future of refrigeration, but for now, the future—as it relates to fridges and the planet in general—is unclear. Heidi Julavits offers this prediction in an essay for the *New Yorker*: "Maybe in the future, Americans will pack our ships, like Tudor did, with ice from our many hotels and general stores and personal ice makers. Maybe, instead of sending our ice to the hotter countries, we'll send it to the previously colder ones."

CHAPTER 14

Flammable Ice

A climate change protest in Union Square, San Francisco (2019)

When Frederic Tudor set out more than two hundred years ago to become the first person to bring Massachusetts ice to the tropics, he did so with the intention of changing how people think of ice. He achieved far more than that. Since that fateful first shipment of ice from Boston Harbor to Martinique, America has become undeniably obsessed with the stuff, and for good reason. Ice has saved lives, reinvented and expanded the food industry, and changed how and where ice sports get played. But while ice has improved lives in so many ways, it has also taken its toll on the planet.

According to an article in *National Geographic*, nearly 8 million tons of plastic waste make their way into oceans every year, choking sea plants and animals and littering coastlines. Plastic ice bags compound the problem. Meanwhile, fridges spew approximately 60 million tons of CO_2 into the atmosphere every year, and those with ice machines, which require additional energy, spew even more. All this CO_2 is causing the planet to warm, and in places where natural ice is still used to store food—such as in Indigenous communities—that warming is rapidly melting the ice they need to survive.

In Utqiaġvik (formerly Barrow), Alaska, the northernmost city in the United States, the Iñupiat people have for centuries stored the whale and fish they hunt for sustenance in ice cellars. Often shared by several families, these cellars drop 10 to 12 feet into the permafrost and can be accessed either through small huts or plywood hatches.

According to an article in *National Geographic*, about "60 percent of Barrow's 1,500 households" rely on shared food, making storage particularly important. Until recently, ice cellars have provided that critical storage. But a warming planet is changing that. Since the 1980s, permafrost temperatures 20 feet below the surface have risen by two or three degrees Fahrenheit. Since the 1950s, average air temperatures have spiked by more than five degrees. As the permafrost melts, it sends water pouring into these underground cellars, flooding them out and spoiling the food inside.

The Iñupiat are adapting as finances allow, purchasing large chest freezers that are delivered to town on cargo ships. Whaling crews, meanwhile, are now sharing walk-in freezer space with scientists at the nearby Naval Arctic Research Laboratory. But what may look to some like economic progress comes at a cultural cost. Ice cellars allow the meat to age slowly, changing its taste to match that which has been enjoyed for centuries—something an electric freezer can't replicate. Moreover, the traditional knowledge that has made ice cellars so successful is at risk of being lost forever.

For non-Indigenous Americans, the use of ice today looks much different than it did in the early nineteenth century, when the tail end of the Little Ice Age froze lakes and rivers to depths made impossible by today's climate. But in some ways, the American *obsession* with ice is much the same. It still reflects the mindset that Frederic Tudor worked hard to instill in the nation: that ice is plentiful, a luxury as well as a necessity—in other words, something the country simply can't live without. Now that humanity finds itself in an age of human-caused global warming, time will tell whether ice machines and polar ice can continue to coexist on the same planet.

In the meantime, a great irony is at play. While the culture that ice created may be contributing to global warming, a very specific kind of ice may help wean Americans—and those in other countries—off fossil fuels. This "savior" ice, as it's been called, is not the same kind of ice that Tudor's men harvested from lakes, that John Gorrie created from machines, or that Shane Truskolaski maintains at the Oval in Utah. This ice is a form of frozen gas called methane hydrate, and it's found throughout the Gulf of Mexico, off the coasts of Japan, and beneath Arctic permafrost. Methane hydrates are essentially balls of methane the size of a human hand encased in lacelike cages of ice that, when touched by a lit match, burst into flame—a property that has earned the substance the nickname "flammable ice."

The science behind these balls of fire and ice is fascinating, but even more exciting is the possibility that they might one day serve as a bridge fuel.

M ethane hydrates are formed through a combination of cold temperatures and relatively high pressure, and are found primarily on the edges of continental shelves, where land slopes sharply into the ocean. They also develop in and beneath permafrost in the Arctic, though they are less prevalent there.

When the pressure that formed them is lowered, or the temperature is raised, they "disassociate," meaning they separate into liquid water and methane gas, which can then, theoretically, be harnessed for energy. And hydrates hold a lot of potential energy. One cubic meter of the material can release up to 160 cubic meters of gas. There is potentially more energy in methane hydrates than in all the world's oil, coal, and gas combined. Moreover, some scientists argue that they are safer to mine than conventional gas or oil.

To gain a better understanding of this fascinating substance, I called Carolyn Ruppel, a research geophysicist at the U.S. Geological

Survey and the project chief for the USGS's Gas Hydrates Project. "There are a few things about hydrates that are interesting," she told me. For one, hydrates form at shallower depths in the ocean than oil, and so are "easier to access than conventional reservoirs" of the fossil fuels that people use now to heat their homes. Another benefit is that gas hydrates come in "a tiny, concentrated package," she said. "So you get much more methane from one of those than from drilling for gas at the same depth." Hydrates also release less carbon dioxide per unit of energy than coal or oil, making them a somewhat cleaner alternative to other fossil fuels. What's more, the BBC has reported that it might be possible to pump CO_2 out of the atmosphere and back into hydrates to replace the methane as it's used. This could "provide an answer to the yet unsolved question of how to store this greenhouse gas safely."

The Japanese government is among the biggest investors in this research, having spent millions of dollars to carry out a number of test projects to determine whether hydrates can be effectively harvested. India, China, and South Korea have also begun to experiment with extraction. In 2000, the United States launched national research and development programs on Alaska's North Slope and offshore in the Gulf of Mexico to determine whether hydrates can be used to replenish or augment domestic fuel supplies. The process of discovery hasn't been smooth, however.

One problem scientists continue to face is how to extract methane hydrates without breaking them down. Extraction lessens the pressure that formed them, causing the hydrates to separate. To avoid this, the gas needs to be extracted from where the hydrates were formed. The Japanese government has funded research to explore this possibility. In 2013, its research team managed to produce gas from the hydrates by drilling into the seabed of the Nankai Trough, which is off the eastern coast of Japan's main island. By lowering pressure on those hydrates, they were able to release and collect the gas for six days before sand entered the well and blocked the supply. A second test completed in 2017 was slightly more successful, running for twenty-four days without technical problems.

These tests suggest that hydrates might one day be a viable energy resource for Japan, but public reaction to the news has been mixed at best. While there has been some excitement about the possibility of Japan becoming energy independent, there has also been some concern that the breakdown of the hydrates could release a lot of methane gas into the ocean and then into the atmosphere, contributing to global warming. And it's not just the Japanese who are worried. Environmentalists from all over the world have expressed concerns. An article in *National Geographic* referred to methane hydrates as "climate time bombs" whose "fuses are burning."

A lot of this worry is unfounded, said Ruppel. Because of the way that water and gas intermingle, "there's not going to be this freight train of methane emitted at the sea floor shooting into the atmosphere." She explained this idea further in a 2017 paper written with John Kessler, a professor in the Department of Earth and Environmental Sciences at the University of Rochester. Most of the methane released from underwater hydrates will oxidize before it ever reaches the surface, never entering the atmosphere. "Unfortunately," she said, "that image [of the runaway freight train] has been promulgated by a lot of people who really push this catastrophism issue. One thing John and I tried to say in [our paper] is, 'Wait a second, guys. If you want to worry about the atmosphere, let's focus on the huge amounts of CO_2 that *humans* are putting in [it].'" Not that the released gases cause zero damage. "To be clear," Ruppel said, "the methane that could be released in the ocean is not innocuous, because it acidifies the ocean—at least a little bit." But even here, the acidification likely to happen from hydrate breakdown is much less than that which is happening because of human activity.

Research is also being done to determine the likelihood that hydrates can be extracted from permafrost, she told me. In collaboration with experts from Japan, the Department of Energy and the U.S. Geological Survey are studying hydrates in Alaska's North Slope. The Arctic environment is quite different from marine environments, but the hydrate reservoirs in both are formed under similar pressure and temperature conditions. Those in the permafrost are just easier to get to.

I mentioned to Ruppel that some environmentalists are concerned that Arctic hydrates will become destabilized as the permafrost thaws, releasing CO_2 into the air and further contributing to global warming. This, too, Ruppel said, is not as great of a danger as some have made it out to be. The hydrate resources there, she said, just aren't that big and their potential contribution to greenhouse gases in the atmosphere is a tiny fraction of the planet's total carbon budget.

When I asked her what, then, the biggest obstructions are to harnessing this energy, she said that first and foremost it's a logistics issue. "Transport is difficult," she said. Not unlike Frederic Tudor, who had to figure out the logistics of shipping ice for long distances at sea, experts still need to determine how best to get gas from the hydrates—or at least their gas emissions—to where they'll be most useful. "For example, if Japan wants to use methane from hydrates," Ruppel told me, "they might have to build a pipeline from land to the Nankai Trough, which is extraordinarily deep and in a place with bad current."

Logistics aren't any easier for the United States, which would have to get hydrates or their gases from the continental shelf or Alaska's permafrost. At the time of this writing, there isn't much incentive to do that. Like Canada, the United States is experiencing a shale gas boom. "The general sense is that, if some countries' backs are up against the wall in twenty years," Ruppel said, "and they would have reasons for not using other sources or those sources are depleted, then these can be a potential backup plan."

Of course, even as a backup plan, methane hydrates aren't completely devoid of environmental issues. While hydrates are somewhat cleaner than other fossil fuels, they are still wrapped up in the same social, economic, and environmental problems as conventional gas and oil, and thus, they will likely only be used as bridge fuel. That means flammable ice may be one of the last new forms of fossil fuels to be extracted on a commercial scale—as well as one of only a few to be developed at a time when the end of fossil fuels is in sight.

Ice may have led to America's obsession with cold, but in the form of methane hydrates, it may help us buy more time on a planet grown too hot.

CONCLUSION

Ice has become an indispensable part of American life. What's more, it's given *shape* to American life; the cultural and economic influences of ice have carved the nation's social contours like a glacier moving across a continent. In late-eighteenth-century New England, ice was a marker of wealth, a sparkling reminder that those with the gold had access to cold. Tudor's brilliant scheme brought ice to places where it rarely—if ever—formed naturally, providing access for people who may not have seen ice before, let alone tasted it. As more merchants joined the trade, the price of ice fell, and more Americans could afford it, changing forever how they stored food, prepared drinks, and practiced medicine. John Gorrie's legacy of mechanical-ice plants lowered prices even further. By the end of the nineteenth century, ice was available to almost everyone in the country.

Fast-forward another century, and bartenders at fancy bars across the United States are making ice extravagant again, at least when it comes to craft cocktails. The last century also witnessed the further proliferation of winter sports, more effective medical treatments, a meteoric rise in electric refrigeration, and, as a result of all this, several million tons of carbon shot into the atmosphere by artificial cooling units like refrigerators and air conditioners.

These are the thoughts that filled my head as I looked at a Rand McNally Road Atlas spread open in front of me on the dining room table. There's something about physical maps I've always loved. They evoke for me a sense of childhood nostalgia the way the iceman does for people fifty years my senior. I flipped the pages to a map of Montana and put my finger on the place I hoped to visit—Glacier National Park.

The park was designated a U.S. national park a little over one hundred years ago. Back then, several of the park's more than 150 mountain peaks featured glaciers—hence its name—which reflected off the area's many alpine lakes. Today, most of those glaciers are greatly diminished or have altogether disappeared. In fact, a recent count suggests that only a couple dozen glaciers remain there, and some of those have lost up to 80 percent of their mass. This is why I wanted to visit the park, and soon—to see the glaciers before they're all gone.

The glaciers are melting because of global warming, and not just in Montana. In Switzerland, glaciers have shrunk by at least 10 percent in the last ten years alone. The Himalayas are often referred to as the third pole because their glaciers are so numerous, but they're rapidly losing ice. The glaciers on these mountains have melted ten times faster in the last four decades than in the previous seven centuries. In 2019, the Icelandic glacier Okjökull lost so much ice, it lost its status as a glacier. The country held a funeral for it—the world's first funeral of this kind—and put a plaque on a boulder near what's left of the ice. The plaque features, in both Icelandic and English, an elegiac message for future generations: "This monument is to acknowledge that we know what is happening and what needs to be done. Only you know if we did it."

The climate crisis we find ourselves in now is not caused solely by the ice trade—far from it. Most of the planet's worst offenders have had nothing to do with ice. But, as we've seen all through this book, the two are entwined in many ways, often with great irony: humans (and Americans in particular) seek out ice to get cool as the planet grows hotter, and as the planet grows hotter, the natural ice that sparked the American appetite for cool continues to disappear.

That's not to say that Americans need to give up their ice obsession (though there are several actions a person can take to live more gently on the earth, including improving their fridge's energy consumption). But such irony does inspire—in me, at least—an appreciation for ice in ways I didn't notice before writing this book. Ice is easy to overlook because it can be purchased now almost anywhere. Indoor ice skating is available in every state in the nation. Ice-cream stands dot the American landscape from ocean to ocean, serving as little oases that can sweeten up long road trips. My own refrigerator has an ice maker in it. I use it regularly to fill a washcloth with ice to nurse a chronic case of shoulder tendonitis. What would I do without ice? I don't want to think of it.

———————

From my home in New England, getting to Glacier National Park will not be easy. The spouse and I could drive, which was why I picked up the atlas. Or we could fly—the closest airport is about thirty miles from the park's west entrance. Neither means of travel is particularly appealing, given their respective carbon outputs. A quick Google search confirms that taking the train is another possibility.

As we consider our options, I feel buoyed by some recent good news. The forecast model that once predicted that the park's glaciers would be completely gone by 2022 thankfully didn't come to pass, and the park's signs that once warned visitors of such a fate have been changed to acknowledge this. The signs now read, "When [the glaciers] will completely disappear depends on how and when we act." The wording is still urgent, but it leaves room for hope.

No one knows for sure what the future will bring, but if the history of ice has taught us anything, it's that entire economic systems and cultural frameworks can be changed in a short amount of time. We've seen what happens when people become so obsessed with ice that they seek new, life-changing ways to sell, create, and consume it. Imagine what would happen if they also seek ways to save it.

ACKNOWLEDGMENTS

A book like this one can exist only because of the work and support of so many talented, generous people. Buckets and buckets of thanks to my agent, Rayhané Sanders, who believed in this book even before I did, and to my editor at Putnam, Michelle Howry, who saw what this book could be and helped me to see the same. Thanks to my intrepid fact-checker, Hilary McClellen, who saved me from my own hubris and ignorance more times than I can count, and to Emma Copley Eisenberg, whose early read and editorial prowess gave me the confidence to finish what I'd started. To the team at Putnam, thank you for treating this book with such care and enthusiasm.

Thanks to the dozens of historians, scientists, and ice experts who answered my endless questions, some of whom appear in this book. To Camper English, whose instructions on how to make better ice have significantly improved my cocktail game, and to Peter and Joann Stack, whose Pennsylvania ice museum should be elevated to one of the wonders of the world—thank you.

The story that this book tells is built on the shoulders of research giants. Thanks to Jonathan Rees, who kindly answered my many questions and whose research into the ice industry and history of refrigeration was invaluable to my own, and to Miranda Rectenwald, librarian at the Special Collections at Washington University in St. Louis, who uncovered vital archival documents related to St. Louis's first ice plants. And thanks to Reinette F. Jones, librarian at the University of Kentucky Special Collections Research Center, whose research into how Black Americans popularized ice cream helped shape this book in profound ways—and will undoubtedly shape the work of many historians to come. And thanks to the many, many other archivists and librarians who hunted down hard-to-find books, letters, diaries, and other materials so that I could at least start to fill the holes of ice history. You are rock stars.

This book is better because of dear friends who graciously agreed to be early readers. Tana Wojczuk, your kind, insightful words buoyed my spirit more times than you know. Thank you for being such a good friend and thoughtful reader. Paul Gagliardi, your hockey expertise helped me to see the game and its history in new ways—thank you, and let's make a toast to your own book sometime soon. And to my gifted friend Tajja Isen, whose own publishing journey is so deeply entangled with mine, thank you for helping to make the writerly process less confusing and a lot more fun.

A more-than-human friend also played an outsize role in helping me write this book. To my late cat Ginsburg, who sat next to me or in my lap as I wrote nearly every word, thank you for your comfort and companionship. I miss you.

To my brilliant colleagues at *Orion*, thank you for your support. What a pleasure and privilege it is to work with you.

Lots of people made sacrifices so that this book could see the light of day, including my parents, in-laws, aunts, and siblings. I'm grateful for all of you. But no one gave more than my life partner and first reader, Alan Scherstuhl, who worked to rearrange our lives so that I could have the time and space to write this book. Thank you, my love, for your endless patience, humor, affection, and support. Life is so much better with you in it.

NOTES

INTRODUCTION

xiii **A 2020 poll conducted** Bosch Home Appliances, "Bosch Study Reveals Americans Are 'Ice Obsessed,' Consuming 400 Pounds Each per Year," *Global Newswire*, December 2, 2020, https://www.globenewswire.com /news-release/2020/12/02/2138421/0/en/Bosch-Study-Reveals-Americans -Are-Ice-Obsessed-Consuming-400-Pounds-Each-Per-Year.html.

xiii **Over 8 million** "Food Storage/Cooking," https://www.energyguide.com /library/EnergyLibraryTopic.asp?bid=austin&prd=10&TID=17257& SubjectID=8371.

xiii **most of which** Daniel Wroclawski, "Best Refrigerators Without a Water Dispenser or Icemaker," *Consumer Reports*, December 7, 2022, https:// www.consumerreports.org/refrigerators/best-refrigerators-without -water-dispenser-or-icemaker-a7916047609.

xvi **Jefferson wrote that the man** "From Thomas Jefferson to Edmund Bacon, 10 November 1806," National Archives, https://founders.archives.gov /?q=Correspondent%3A%22Jefferson%2C%20Thomas%22%20 Correspondent%3A%22Bacon%2C%20Edmund%22&s=1111311111&r=5.

xvi **he gawked at the American** Charles Dickens, *American Notes for General Circulation and Pictures from Italy* (London: Chapman and Hall, 1913), https://www.gutenberg.org/files/675/675-h/675-h.htm.

xvi **Ice was so prominent** Ben Kemp, "A Necessary Undertaking," Grant Cottage National Historic Landmark, July 29, 2019, https://www .grantcottage.org/blog/2019/7/29/a-necessary-undertaking.

xvii **The Chinese were cooling** Zheng Songwu, "How Did the Ancient Chinese Stay Cool in Summer?," *CGTN*, August 14, 2018, https://news.cgtn.com /news/3d3d414d3167444e79457a6333566d54/share_p.html.

xvii   **The Persians were building** "Yakhchal—Ancient Type of Refrigerator,"
        History of Refrigeration, http://www.historyofrefrigeration.com
        /refrigeration-history/yakhchal-ancient-refrigerator/.

xvii   **The British built their** "The Ice House Uncovered," *Country Life*, October
        4, 2010, https://www.countrylife.co.uk/comment-opinion/the-ice-house
        -uncovered-20789.

xvii   **The Iñupiat have for centuries** Andy Gregory and Rachel d'Oro, "'I'm
        Worried': Alaska's Ice Cellars Melting Due to Climate Change after Being
        Used to Store Food for Generations," *Independent*, November 25, 2019, https://
        www.independent.co.uk/climate-change/news/climate-change-permafrost
        -alaska-ice-cellars-melting-inupiat-food-storage-whaling-a9216571.html.

xvii   **Ice is so critical** Oliver Milman, "Alaska Indigenous People See Culture
        Slipping Away as Sea Ice Vanishes," *Guardian*, December 19, 2016, https://
        www.theguardian.com/environment/2016/dec/19/alaska-sea-ice
        -vanishing-climate-change-indigenous-people.

xx   **"The American need for ice"** Heidi Julavits, "American Exceptionalism on
        Ice," *New Yorker*, July 8, 2016, https://www.newyorker.com/culture
        /culture-desk/american-exceptionalism-on-ice.

CHAPTER 1 · THE MAN WHO WOULD BE ICE KING

6   **In February 1801** Gavin Weightman, *The Frozen Water Trade: A True Story*
        (New York: Hyperion, 2003), 37.

7   **The family chef had told** Weightman, *The Frozen Water Trade*, 20.

7   **As the ship departed** Alan Seaburg, ed., *The Ice King: Frederic Tudor and
        His Circle* (Boston: Massachusetts Historical Society, 2003), 12.

7   **downpour flooded the deck** Seaburg, *The Ice King*, 12.

7   **The sun came out** Seaburg, *The Ice King*, 12.

7   **the brothers death-marched** Weightman, *The Frozen Water Trade*, 18.

7   **Frederic hired an English-speaking** Seaburg, *The Ice King*, 12.

8   **if he had "to fight"** John Tudor, "Journal of a Tour in Search of Health,"
        April 10, 1801, Papers of John Henry Tudor, bMS Am 1197 and bMS Am
        2027, Houghton Library, Harvard University, Cambridge, Massachusetts.

8   **John Henry's knee** Seaburg, *The Ice King*, 13.

8   **The insect invasion** Jim Murphy, *An American Plague: The True and
        Terrifying Story of the Yellow Fever Epidemic of 1793* (New York: Clarion
        Books, 2003), 99.

8   **People throughout Havana** Murphy, *An American Plague*, 100–104.

8   **Their luck didn't improve** Weightman, *The Frozen Water Trade*, 18–19.

8   **John Henry's health** Seaburg, *The Ice King*, 14.

9   **His death shook Frederic** Little has been written about Frederic Tudor's relationship with his brother, but diary entries at this time suggest the tremendous emotional toll John Henry's death took on him.

9   **His family and friends** Frederic Tudor, diary, August 1805, Tudor Company Records, Baker Library, Harvard Business School, Boston, Massachusetts.

9   **"He who gives back"** Tudor, diary, December 1805.

9   **"in the warm season"** "Ice House," George Washington's Mount Vernon, https://www.mountvernon.org/the-estate-gardens/location/ice-house/.

9   **Every winter, the Tudor** Weightman, *The Frozen Water Trade*, 45.

10   **If packed tightly** For an in-depth examination of the labor involved in the ice harvest, see Jonathan Rees, *Before the Refrigerator: How We Used to Get Ice* (Baltimore: Johns Hopkins University Press, 2018), especially pages 25–27 and 34.

10   **"Some enterprising Yankee"** Tudor, diary, November 1805.

11   **"Our plan now"** Tudor, diary, November 1805.

11   **In the following weeks** Frederic Tudor to Harrison Gray Otis, November 1805, Tudor Company Records, Baker Library, Harvard Business School, Boston, Massachusetts.

11   **There's nothing in** Bodil Bjerkvik Blain, "Melting Markets: The Rise and Decline of the Anglo-Norwegian Ice Trade, 1850–1920," Working Papers of the Global Economic History Network (GEHN), no. 20/06 (February 2006).

11   **The senator refused** Tudor, diary, November 1805.

11   **Frederic confined his annoyance** Tudor, diary, November 1805.

11   **In a letter** Tudor, diary, November 1805.

12   **He was a poet** *North American Review*, https://northamericanreview.org/history.

12   **William recruited their cousin** Seaburg, *The Ice King,* 26.

12   **On a warmer-than-usual** Weightman, *The Frozen Water Trade*, 27.

12   **When the *Jane*** Weightman, *The Frozen Water Trade*, 33.

12   **The night they approached** Weightman, *The Frozen Water Trade*, 28.

13   **But this was no help** Sue Peabody, "French Emancipation," *Oxford Bibliographies*, October 28, 2014, https://www.oxfordbibliographies.com/view/document/obo-9780199730414/obo-9780199730414-0253.xml.

13   **By the time** Weightman, *The Frozen Water Trade*, 25

13   **local prefect granted** Weightman, *The Frozen Water Trade*, 35, 41.

NOTES

13   **This stretch of the trip** Seaburg, *The Ice King*, 29.

14   **They managed to find** Seaburg, *The Ice King*, 29.

14   **"A seaman's life"** James Savage to Elizabeth Savage, June 1, 1806, Ms. N-231, James Savage Papers II, Massachusetts Historical Society, Boston, Massachusetts.

14   **"The advantages derived"** Tudor, diary, 1806.

14   **When he calmed down** Weightman, *The Frozen Water Trade*, 29.

15   **The ship alone cost** Weightman, *The Frozen Water Trade*, 29.

15   **Even as their plan** Seaburg, *The Ice King*, 30.

15   **"People only laugh"** Tudor, diary, 1805.

16   **Not until two weeks** Weightman, *The Frozen Water Trade*, 32.

16   **"A Vessel has cleared"** *Boston Gazette*, 1806, Boston Gazette Commercial and Political, Library of Congress, Washington, D.C.

16   **Only now did Frederic** Weightman, *The Frozen Water Trade*, 34.

17   **He charged sixteen** Weightman, *The Frozen Water Trade*, 33.

17   **The islander gestured** Seaburg, *The Ice King*, 32.

17   **Toward the end** Seaburg, *The Ice King*, 32.

18   **He sought new** Weightman, *The Frozen Water Trade*, 45.

18   **The capital city reacted** Weightman, *The Frozen Water Trade*, 45.

18   **Just when Frederic thought** Weightman, *The Frozen Water Trade*, 46.

18   **Frederic was mortified** Weightman, *The Frozen Water Trade*, 46.

18   **By 1809, Frederic** Weightman, *The Frozen Water Trade*, 55.

18   **he was again arrested** Weightman, *The Frozen Water Trade*, 48.

19   **It took Frederic** Weightman, *The Frozen Water Trade*, 50.

19   **The day he arrived** Weightman, *The Frozen Water Trade*, 59.

19   **Part one of** Weightman, *The Frozen Water Trade*, 60.

20   **For the first time** Weightman, *The Frozen Water Trade*, 60.

20   **When he returned** Weightman, *The Frozen Water Trade*, 60.

20   **Demand in Cuba kept** Hamish Anderson, "Inside Cuba's Intense Ice Cream Obsession," *Saveur*, June 29, 2016.

20   **Castro was famously** Richard Boudreaux, "Castro's Revolutionary Cry: Let Them Eat Ice Cream!," *Los Angeles Times*, November 5, 1991.

20   **novelist Gabriel García Márquez** Jennifer Simonson, "The Curious History of Cuba's Ice Cream Obsession," *Telegraph*, July 8, 2019, https://www.telegraph.co.uk/travel/destinations/caribbean/cuba/articles /cuba-ice-cream-fidel-castro-coppelia/.

20   **Castro took up the slogan** Nancy Stout, *One Day in December: Celia Sánchez and the Cuban Revolution* (New York: Monthly Review Press, 2013), 194.

21   **Cuban originals such as** Jason Motlagh, "The Future of Cuba's Socialist Ice-Cream Cathedral," *The Guardian*, April 14, 2015, https://www.theguardian.com/world/2015/apr/14/future-of-coppelia-cuba-socialist-ice-cream-cathedral.

21   **Castro's parlor still sells** Motlagh, "The Future of Cuba's Socialist Ice-Cream Cathedral."

21   **Grateful to be taken seriously** Seaburg, *The Ice King*, 91.

21   **All was going well** Weightman, *The Frozen Water Trade*, 90.

22   **"It is absolutely necessary"** Weightman, *The Frozen Water Trade*, 90.

22   **a ship full of Frederic's ice set sail** Seaburg, *The Ice King*, 92.

22   **He would have found them in** Joan Garvey, *Beautiful Crescent: A History of New Orleans* (Gretna, LA: Pelican Publishing, 2013), 55, 90.

23   **Drawing on Creole and Spanish** Elizabeth M. Williams and Chris McMillian, *Lift Your Spirits: A Celebratory History of Cocktail Culture in New Orleans* (Baton Rouge: Louisiana State University Press, 2016), 12–14.

23   **The Tudors' go-to recipe** Weightman, *The Frozen Water Trade*, 147.

24   **Bartenders soon discovered** Williams and McMillian, *Lift Your Spirits*, 45.

24   **New Orleans mixologists** Williams and McMillian, *Lift Your Spirits*, 49.

24   **By 1840, fleets** Weightman, *The Frozen Water Trade*, 132.

24   **He worked with a former** Weightman, *The Frozen Water Trade*, 87.

25   **They stored their ice** Lydia Bjornlund, *How the Refrigerator Changed History* (North Mankato, MN: Essential Library, 2015), 27.

25   **The icebox lowered** Charlette Gallagher and John Allred, *Taking the Fear Out of Eating* (Cambridge: Cambridge University Press, 1992), 144.

25   **"Ice must be considered"** Tudor, diary, 1805.

25   **American beer industry** "National Beer Sales and Production Data," Brewers Association, https://www.brewersassociation.org/statistics-and-data/national-beer-stats/.

25   **The industry wouldn't have existed** Susan K. Appel, "Artificial Refrigeration and the Architecture of 19th-Century American Breweries," *Journal of the Society for Industrial Archeology* 16, no. 1 (1990): 21.

25   **Made from a fermentation** Appel, "Artificial Refrigeration," 22–24.

26   **In 1844, the Pabst Brewing Company** Lee E. Lawrence, "The Wisconsin Ice Trade," *Wisconsin Magazine of History* 48, no. 4 (Summer 1965): 258.

26   **For years, the Great Lakes** Weightman, *The Frozen Water Trade*, 190.

26   **The plan was ruthless** Lawrence, "The Wisconsin Ice Trade," 261–64.

26   **In a clandestine visit** Lawrence, "The Wisconsin Ice Trade," 264.

27   **The company kept** Lawrence, "The Wisconsin Ice Trade," 264.

27   **Pike Co. distributed** Weightman, *The Frozen Water Trade*, 190.

27   **The band kept playing** Weightman, *The Frozen Water Trade*, 190.

27   **rivals became colleagues** Weightman, *The Frozen Water Trade*, 190.

27   **"charming twenty-nine-year-old widow"** Seaburg, *The Ice King*, 124.

28   **met her at a ball** Seaburg, *The Ice King*, 124.

28   **he received a package** Seaburg, *The Ice King*, 125.

28   **"Free life, no marriage"** Tudor, diary, December 1930.

28   **Euphemia was pregnant** Seaburg, *The Ice King*, 175.

28   **"delicate troubles"** Seaburg, *The Ice King*, 175.

28   **The second doctor** Seaburg, *The Ice King*, 175.

28   **"flamed like a comet"** Seaburg, *The Ice King*, 203–4.

28   **Another story tells** Seaburg, *The Ice King*, 203.

29   **"I am 64 years"** Frederic Tudor to Euphemia F. Tudor, December 7, 1847, Tudor Company Records, Baker Library, Harvard Business School, Boston, Massachusetts.

29   **After his death** Seaburg, *The Ice King*, 207.

29   **"Truth," she wrote** Tudor, diary, 1863.

CHAPTER 2   ·   BLASPHEMOUS ICE

33   **nearly 750,000 Americans** Guy Gugliotta, "New Estimate Raises Civil War Death Toll," *New York Times*, April 2, 2012, https://www.nytimes.com /2012/04/03/science/civil-war-toll-up-by-20-percent-in-new-estimate .html.

33   **Without ice, southern hospitals** Ira Rutkow, *Bleeding Blue and Gray: Civil War Surgery and the Evolution of American Medicine* (New York: Stackpole Books, 2015), 42–45.

34   **Without it, soldiers** Rutkow, *Bleeding Blue and Gray*, 45.

34   **"Sick soldiers in Augusta"** James Ford Rhodes, *History of the United States from the Compromise of 1850*, vol. 1: *1850–1854* (New York: Macmillan, 1910), 351.

36   **To earn money** V. M. Sherlock, *The Fever Man: A Biography of Dr. John Gorrie* (Apalachicola, FL: Medallion Press, 1982), 5.

NOTES

36   **Yellow fever killed** Linda Caldwell, *He Made Ice and Changed the World: The Story of Florida's John Gorrie* (Ocala, FL: Atlantic Publishing Group, 2019), 28.

36   **During fever season's peak** Elli Morris, *Cooling the South: The Block Ice Era, 1875–1975* (Richmond, VA: Wackophoto, 2008), 38.

37   **left patients feeling dizzy** Sherlock, *The Fever Man*, 6–7.

37   **Even a splash** James D. Lutz, "Lest We Forget, a Short History of Housing in the United States," American Council for an Energy-Efficient Economy, https://www.aceee.org/files/proceedings/2004/data/papers/SS04 _Panel1_Paper17.pdf.

37   **The following spring** Sherlock, *The Fever Man*, 7.

37   **Gorrie pondered where** Sherlock, *The Fever Man*, 40.

38   **When he arrived** Salvatore Basile, *Cool: How Air Conditioning Changed Everything* (New York: Fordham University Press, 2016), 22.

38   **So were violent clashes** Michael Gannon, *The History of Florida* (Gainesville: University Press of Florida, 2013), 45–48.

38   **bringing a wave** Morris, *Cooling the South*, 38.

38   **He soon had** Caldwell, *He Made Ice and Changed the World*, 18.

39   **The thunderous shots** Caldwell, *He Made Ice and Changed the World*, 62.

39   **believed that crowd poison** Basile, *Cool*, 15.

39   **Stages and pulpits** Basile, *Cool*, 5–8.

39   **"a two hours seething"** Thomas Hood, "Hot Weather at the Play," *The Works of Thomas Hood* (London: E. Moxon, Son and Co., 1873).

39   **Snowstorms were thought** Basile, *Cool*, 9.

40   **Ice was dubbed** Basile, *Cool*, 25.

40   **When shipments arrived** Caldwell, *He Made Ice and Changed the World*, 93.

40   **The next month brought** Caldwell, *He Made Ice and Changed the World*, 16–17.

40   **yellow fever took** Jason Dehart, "Yellow Fever Was the Scourge of Tallahassee and Surrounding Towns in 1841," *Tallahassee Magazine*, June 25, 2012, https://www.tallahasseemagazine.com/yellow-fever-was -the-scourge-of-tallahassee-and-surrounding-towns-in-1841/.

41   **Within minutes, the room** Caldwell, *He Made Ice and Changed the World*, 93.

41   **Somehow, he maintained** Sherlock, *The Fever Man*, 53. See also Caldwell, *He Made Ice and Changed the World*, 41–42.

42   **The ice-making prototype** Caldwell, *He Made Ice and Changed the World*, 97–99.

42   **capable of creating** Caldwell, *He Made Ice and Changed the World*, 102.

42   **Gorrie had ideas** Caldwell, *He Made Ice and Changed the World*, 99.

42 **Gorrie saw Chapman** Sherlock, *The Fever Man*, 103.

42 **His reaction was less** Sherlock, *The Fever Man*, 103–5.

43 **his first public admissions** Caldwell, *He Made Ice and Changed the World*, 89.

43 **"We do not know"** Quoted in Basile, *Cool*, 22.

43 **"cock-and-bull story"** Quoted in Basile, *Cool*, 26.

43 **Hiding behind his** Caldwell, *He Made Ice and Changed the World*, 89.

43 **Gorrie continued to write** Caldwell, *He Made Ice and Changed the World*, 55.

43 **"the most destructive system"** Caldwell, *He Made Ice and Changed the World*, 55.

44 **With the last of** Basile, *Cool,* 26.

44 **He approached investors** Basile, *Cool,* 26.

44 **willing publicity partner** Basile, *Cool,* 26.

45 **he could demonstrate** Sherlock, *The Fever Man*, 105–6.

45 **"On Bastille Day"** Minna Scherlinder Morse, "Chilly Reception," *Smithsonian Magazine*, July 2002, 41–42.

45 **carrying small blocks** Morse, "Chilly Reception," 42.

45 **the consul's party guests** Basile, *Cool,* 26.

46 **"There is a crank"** Basile, *Cool, 26.*

46 **Gorrie was astonished** Basile, *Cool,* 26.

46 **secured a U.S. patent** Caldwell, *He Made Ice and Changed the World*, 99.

46 **found an investor** Basile, *Cool,* 27.

46 **hawking pamphlets about his idea** Basile, *Cool,* 27.

46 **Gorrie sat shivering** Basile, *Cool,* 27.

47 **"has been found"** Basile, *Cool,* 27.

47 **Free from Tudor's reach** Basile, *Cool,* 35–36.

47 **The South in particular** Elli Morris, "Making Ice in Mississippi," *Mississippi History Now,* May 2010, https://www.mshistorynow.mdah.ms.gov/issue/making-ice-in-mississippi.

48 **bedpan "cooling machine"** Caldwell, *He Made Ice and Changed the World*, 114.

48 **machine's popularity peaked** Ira Rutkow, *James A. Garfield: The American Presidents Series; The 20th President, 1881* (New York: Times Books, 2006), 25.

48 **Merchants from those countries** Jonathan Rees, *Refrigeration Nation: A History of Ice, Appliances, and Enterprise in America* (Baltimore: Johns Hopkins University Press, 2013), 82.

48 **pack their cargo hulls** Rees, *Refrigeration Nation*, 91–92.

49   **ice blocks were placed** Will Oremus, "A History of Air Conditioning," *Slate*, July 15, 2013, https://slate.com/culture/2013/07/a-history-of -air-conditioning.html. The word "block buster," and later "blockbuster," has multiple origins. While the use of ice and rudimentary air conditioning likely contributed to its use, it didn't become the ubiquitous term it is today until the mid-twentieth century, when Time magazine published an article in 1942 about the British Royal Air Force's bombs that could destroy entire city blocks. These bombs became known as blockbusters.

CHAPTER 3 · THE ICEMAN COMETH

54   **One recurring gag** Christopher Miller, *American Cornball: A Laffopedic Guide to the Formerly Funny* (New York: HarperCollins, 2014), 244.

54   **"[Icemen] were symbols"** Jonathan Rees, *Before the Refrigerator: How We Used to Get Ice* (Baltimore: Johns Hopkins University Press, 2018), 77.

54   **work was strenuous** Rees, *Before the Refrigerator*, 77.

55   **Perhaps history's most famous** Robert S. Gallagher, "The Galloping Ghost: An Interview with Red Grange," *American Heritage* 26, no. 1 (December 1974).

55   **"unlike the milkman"** Miller, *American Cornball*, 243.

55   **"Thus, his ethnicity"** Miller, *American Cornball*, 243.

55   **approximately three thousand horses** Rees, *Before the Refrigerator*, 76.

56   **Ice wagons working** Rees, *Before the Refrigerator*, 76.

56   **bright colors faded** Rees, *Before the Refrigerator*, 76.

56   **Wagons carrying natural ice** Rees, *Before the Refrigerator*, 57.

56   **It wasn't uncommon** Rees, *Before the Refrigerator*, 76.

57   **Coupon books were distributed** Rees, *Before the Refrigerator*, 83.

57   **commercial truck drivers** "Trucking Industry History and Overview," *The Trucker*, accessed December 26, 2022, https://www.thetrucker.com /truck-driving-jobs/resources/trucking-industry-history-and-overview.

57   **Toxic waste from** Jonathan Rees, *Refrigeration Nation: A History of Ice, Appliances, and Enterprise in America* (Baltimore: John Hopkins University Press, 2013), 55.

57   **hundreds of cattle** John L. Puckett, "Lovely Abattoir on the Schuylkill: The Railroad and the Stock-Yard Company," West Philadelphia Collaborative History, accessed December 26, 2022, https:// collaborativehistory.gse.upenn.edu/stories/lovely-abattoir-schuylkill -railroad-and-stock-yard-company.

57   **"a dark greenish color"** Rees, *Refrigeration Nation,* 55.

57   **"not a proper water"** Rees, *Refrigeration Nation*, 55.

57 got around these restrictions Rees, *Refrigeration Nation*, 55.

58 Germ theory was still new Gavin Weightman, *The Frozen Water Trade: A True Story* (New York: Hyperion, 2003), 195.

58 The hospital suffered Rees, *Refrigeration Nation*, 62.

58 "It is probable" Rees, *Refrigeration Nation*, 62.

58 In Chicago, a similar Rees, *Refrigeration Nation*, 59.

59 "We furnish PURE EASTERN" Rees, *Refrigeration Nation*, 56.

59 The only penalty Rees, *Refrigeration Nation*, 61.

59 "could be seen" Rees, *Refrigeration Nation*, 60.

59 "The character of the lake" Rees, *Refrigeration Nation*, 74.

59 "The difference between" Rees, *Refrigeration Nation*, 64.

59 "No sewage" Rees, *Refrigeration Nation*, 73.

60 such ice accounted David Hemenway, "The Ice Trust," in *Prices and Choices: Microeconomic Vignettes*, ed. David Hemenway (Lanham, MD: University Press of America, 1993), 190.

60 "The time has arrived" Rees, *Refrigeration Nation*, 69–70.

60 *Washington Evening Star* reported "Supply of Ice is a Necessity," *Washington Evening Star*, August 12, 1906, 3.

61 running sixty barges Rees, *Before the Refrigerator*, 76.

61 Morse bought up smaller Rees, *Before the Refrigerator*, 42.

61 Morse hiked up prices Rees, *Before the Refrigerator*, 42.

62 "unless something is done" Rees, *Before the Refrigerator*, 41.

62 deadly heat wave struck Edward P. Kohn, *Hot Time in the Old Town: The Great Heat Wave of 1896 and the Making of Theodore Roosevelt* (New York: Basic Books, 2010), 5.

62 became ovens with temperatures Kohn, *Hot Time in the Old Town*, 6.

62 "The sidewalks were lined" Kohn, *Hot Time in the Old Town*, 116.

62 "It is not the providence" Kohn, *Hot Time in the Old Town*, xii.

62 By the time the heat Kohn, *Hot Time in the Old Town*, x.

63 giving meteoric rise Kohn, *Hot Time in the Old Town*, xii.

63 "the worst and most fatal" Kohn, *Hot Time in the Old Town*, ii.

63 Morse continued to build Hemenway, "The Ice Trust," 190.

63 He now owned virtually Hemenway, "The Ice Trust," 190.

64 He obtained a permanent Hemenway, "The Ice Trust," 191.

64 bought-out ice dealer Hemenway, "The Ice Trust," 191.

NOTES

64   **proved that the ice shortages** Hemenway, "The Ice Trust," 192.

64   **"ONE HUNDRED PER CENT RISE"** Rees, *Before the Refrigerator*, 42.

64   **despite the price hike** Suzanne Spellen, "Walkabout: It's Nice to Be Ice, Part 3," *Brownstoner*, January 9, 2014, https://www.brownstoner.com /history/walkabout-its-nice-to-be-ice-part-three/.

65   **dropped another bombshell** Spellen, "Walkabout: It's Nice to Be Ice, Part 3."

65   **destroying the careers** Hemenway, "The Ice Trust," 194.

65   **lost his run for governor** Hemenway, "The Ice Trust," 193.

65   **"is talked of everywhere"** Hemenway, "The Ice Trust," 194.

66   **averse only to "bad" trusts** Hemenway, "The Ice Trust," 195.

66   **largest facility, in Iceboro** Suzanne Spellen, "Walkabout: It's Nice to Be Ice, Part 4," *Brownstoner*, January 14, 2014, https://www.brownstoner.com /history/walkabout-its-nice-to-be-ice-part-four/.

66   **One year later** Elliot Carter, "American Ice Co.," Architect of the Capitol, August 5, 2016, https://architectofthecapital.org/posts/2016/8/5 /american-ice-co.

66   **Morse had silently withdrawn** Hemenway, "The Ice Trust," 197.

66   **build another monopoly** Hemenway, "The Ice Trust," 197.

66   **Morse wasn't finished** Hemenway, "The Ice Trust," 197.

67   **The banks were forced** Robert F. Bruner and Sean D. Carr, *The Panic of 1907: Lessons Learned from the Market's Perfect Storm* (Hoboken, NJ: John Wiley and Sons, 2007), 32.

67   **"fat, squatty little man"** Hemenway, "The Ice Trust," 197.

67   **"There is no one"** Hemenway, "The Ice Trust," 197.

67   **Morse was sentenced** Mitchell Zuckoff, *Ponzi's Scheme: The True Story of a Financial Legend* (New York: Random House, 2005), 274.

67   **While there he umpired** Doug Bertlesman, "The Ice King of Bath," *Meander Maine*, accessed December 26, 2022, https://meandermaine.com /tale/the-ice-king/.

67   **friends and relatives attempted** "Charles W. Morse," *Dictionary of American Biography*, vol. 8 (London: Forgotten Books, 2018), 240–41.

68   **Given Morse's symptoms** "Charles W. Morse," *Dictionary of American Biography*, 240–41.

68   **Morse had drunk** "Charles W. Morse," *Dictionary of American Biography*, 240–41.

68   **"shakes one's faith"** "Charles W. Morse," *Dictionary of American Biography*, 240–41.

68   **once his legal danger passed** Hemenway, "The Ice Trust," 197.

69   **These new ice machines** Rees, *Refrigeration Nation*, 74.

69   **cut off the supply** Sarah Wassberg Johnson, "World War Wednesdays: Ice Is Needed," *The Food Historian*, December 4, 2019, https://www .thefoodhistorian.com/blog/world-war-wednesdays-ice-is-needed.

69   **"help in the harvest"** United States Food and Drug Administration, poster, National Association of Ice Industries archive, Library of Congress, Washington, D.C.

69   **Ice deliverywomen, often working** "Behind the Picture: Girls Delivering Ice (1918)," *Lomography Magazine*, accessed December 26, 2022, https:// www.lomography.com/magazine/314806-behind-the-picture-girls -delivering-ice-1918.

70   **Returning-soldier husbands** "The History of Ice and the IPIA," video, International Packaged Ice Association, accessed December 26, 2022, https://www.packagedice.com/history-of-ice.html.

70   **Within a few years** Gina Medsker, "Refrigerators: 1940s–1950s," Evolution of Home Appliances in the US, accessed December 26, 2022, https:// evolutionhomeappliances.weebly.com/refrigerators-1940s-1950s-new -features-and-the-deep-freeze.html.

CHAPTER 4 · HIGHWAYS, HOLIDAYS, AND THE CHEAP-ICE REVOLUTION

74   **Many of them went** Nathan Miller, *New World Coming: The 1920s and the Making of Modern America* (New York: Scribner, 2010), 182.

74   **described by his colleagues** Allen Liles, *Oh Thank Heaven! The Story of the Southland Corporation* (Texas: Southland Corporation, 1977), 16.

75   **taught him that spectacle sells** Liles, *Oh Thank Heaven!*, 16.

75   **Green noticed that many** "Red Circle and Gold Leaf," *Time*, November 13, 1950, https://content.time.com/time/subscriber/article /0,33009,821397,00.html.

75   **His strategy worked** Liles, *Oh Thank Heaven!*, 20.

75   **Green's hours "sacrilege"** Liles, *Oh Thank Heaven!*, 20.

75   **provide customers with** Liles, *Oh Thank Heaven!*, 23.

76   **creating the first outline** Liles, *Oh Thank Heaven!*, 27.

76   **mass-producing refrigerators** Mary Cross, *Century of American Icons: 100 Products and Slogans from the 20th-Century Consumer Culture* (Westport, CT: Greenwood Press, 2002), 54–55.

76   **sending their old iceboxes** Jonathan Rees, *Refrigeration Nation: A History of Ice, Appliances, and Enterprise in America* (Baltimore: Johns Hopkins University Press, 2013), 137.

76 **the woman-to-woman approach** Liles, *Oh Thank Heaven!*, 27.

76 **In the 1930s** Rees, *Refrigeration Nation*, 166.

77 **rural homes began** Rees, *Refrigeration Nation*, 166.

77 **Soldier camps required** Liles, *Oh Thank Heaven!*, 57.

77 **"did not have a lot"** Liles, *Oh Thank Heaven!*, 57.

77 **"about a hundred 300-pound blocks"** Liles, *Oh Thank Heaven!*, 58.

78 **"It was a 60-mile"** Liles, *Oh Thank Heaven!*, 58.

78 **By 1942, ice plants** Elli Morris, "Making Ice in Mississippi," *Mississippi History Now,* May 2010, https://www.mshistorynow.mdah.ms.gov/issue /making-ice-in-mississippi.

78 **By the time the war** Liles, *Oh Thank Heaven!*, 60–62.

78 **"America's Handsomest Iceman Contest"** "The History of Ice and the IPIA," video, International Packaged Ice Association, accessed December 26, 2022, https://www.packagedice.com/history-of-ice.html.

79 **savvier fast-food restaurants saw** Helen Rosner, "Pellet Ice Is the Good Ice," *New Yorker*, January 27, 2021.

79 **No other store model** Liles, *Oh Thank Heaven!*, 67.

80 **other companies modeled** Steven M. Graves, "Convenience Stores: A Landscape Perspective," *Yearbook of the Association of Pacific Coast Geographers* 79 (2017): 134–52.

80 **convenience mattered more** Graves, "Convenience Stores."

80 **farmer from Barnes** Beccy Tanner, "Coffeyville Man Creates ICEE," *Wichita Eagle*, April 16, 2012.

80 **On especially hot** Ron Wilson, "Remembering Omar Knedlik, the Rural Kansas Man Who Invented the ICEE," Kansas Public Radio, September 25, 2017, https://kansaspublicradio.org/kpr-news /remembering-omar-knedlick-kansas-man-who-invented-icee.

80 **7-Eleven purchased** Tanner, "Coffeyville Man Creates ICEE."

81 **Ads for cars helped** Peter J. Blodgett, "How Americans Fell in Love with Taking Road Trips," *Time*, August 15, 2015, https://time.com/3998949 /road-trip-history.

81 **"He has no beaten track"** Blodgett, "How Americans Fell in Love with Taking Road Trips."

81 **While traveling by car** Gina Medsker, "Refrigerators: 1940s–1950s," Evolution of Home Appliances in the US, https://evolutionhomeappliances.weebly.com /refrigerators-1940s-1950s-new-features-and-the-deep-freeze.html.

81 **first "portable ice chest"** United States Patent Office, *Official Gazette of the United States Patent Office* 677, no. 1 (December 1, 1953), 931.

82   It would be another decade Graves, "Convenience Stores."

82   Among the many Americans "Rapid Rise of the Host with the Most,"
        Time, June 12, 1972, https://content.time.com/time/subscriber/article
        /0,33009,906054-1,00.html.

82   the chain boasted "Rapid Rise of the Host with the Most," *Time*.

CHAPTER 5 · THE INSIDE SCOOP

92   "they may give way" "Ice, a Most Luxurious Crop," George Washington's
        Mount Vernon, https://www.mountvernon.org/the-estate-gardens/ice/.

92   "The large number of ice-cream" Mary V. Thompson, MA, DLitt, research
        historian at the Fred W. Smith National Library for the Study of George
        Washington, Mount Vernon Estate and Gardens, to John L. Smith Jr.,
        February 21, 2014.

92   most families did not have Smith, "Washington: Father of Two Scoops."

93   The Washingtons were likely Louise Conway Belden, *The Festive
        Tradition: Table Decoration and Desserts in America, 1650–1900* (New York:
        W. W. Norton, 1983), 145.

93   "Cream Machine for Ice" Stephen Decatur Jr., *Private Affairs of George
        Washington: From the Records and Accounts of Tobias Lear, Esquire, His
        Secretary* (Boston: Houghton Mifflin, 1933), 253.

93   George and Martha popularized Eugene E. Prussing, *The Estate of George
        Washington, Deceased* (Boston: Little, Brown, 1927), 435.

93   But the tale's persistence Marilyn Powell, *Ice Cream: The Delicious History*
        (New York: Abrams Press, 2009): 11.

93   "people spin all sorts" Jeri Quinzio, *Of Sugar and Snow: A History of Ice
        Cream Making* (Berkeley: University of California Press, 2009), ix.

93   "there would be references" Quinzio, *Of Sugar and Snow*, ix–x.

94   need to balance sugar Quinzio, *Of Sugar and Snow*, 6–9.

94   "strange and decadent" Anne Cooper Funderburg, *Chocolate, Strawberry,
        and Vanilla: A History of American Ice Cream* (Bowling Green, OH: Bowling
        Green State University Popular Press, 1995), 3.

94   In Rozzano, he Funderburg, *Chocolate, Strawberry, and Vanilla*, 16.

94   "snow gives the most delicate" Quinzio, *Of Sugar and Snow*, 76.

94   His famous recipe Funderburg, *Chocolate, Strawberry, and Vanilla*, 16.

94   popularized ice cream Funderburg, *Chocolate, Strawberry, and Vanilla*, 17.

95   "Pastry and pudding going out" Josephine Seaton, *William Winston
        Seaton of the "National Intelligencer": A Biographical Sketch* (Boston: James
        R. Osgood, 1871), 84–88.

95   "ice-cream, put in a silver dish" Mary Boardman Crowninshield, *Letters of Mary Boardman Crowninshield, 1815–1816* (Cambridge, MA: Riverside Press, 1935), 25–26.

95   "It is the practice with some" Funderburg, *Chocolate, Strawberry, and Vanilla*, 31–32.

96   first American shops to sell Quinzio, *Of Sugar and Snow*, 80–81.

96   That's one reason Maria Panaritis, "The Long-Lost 'Father of Ice Cream,'" *Philadelphia Inquirer*, August 4, 2019, accessed August 21, 2022, https://www.newspapers.com/image/legacy/592310685.

97   Jackson's ice creams created Funderburg, *Chocolate, Strawberry, and Vanilla*, 14.

97   considered to be too rustic Quinzio, *Of Sugar and Snow*, 81.

98   "a quarter slyly dropped" Quinzio, *Of Sugar and Snow*, 82–83.

98   His establishment included Quinzio, *Of Sugar and Snow*, 90.

98   The democratization of pleasure gardens Quinzio, *Of Sugar and Snow*, 84.

98   "to define and establish" Naomi J. Stubbs, *Cultivating National Identity through Performance: American Pleasure Gardens and Entertainment* (New York: Palgrave Macmillan, 2013), 77.

98   A handful of newspapers Quinzio, *Of Sugar and Snow*, 84.

99   A local newspaper Quinzio, *Of Sugar and Snow*, 83.

99   On his theater's Marvin McAllister, *White People Do Not Know How to Behave at Entertainments Designed for Ladies and Gentlemen of Colour: William Brown's African and American Theater* (Chapel Hill, NC: University of North Carolina Press, 2003), 3.

99   Collot was most familiar Reinette F. Jones, "Ice Cream, Sugar and Slavery up to the 1900s," Notable Kentucky African Americans Database, March 19, 2020 https://nkaa.uky.edu/nkaa/items/show/300003935.

100   In Kentucky, Mrs. Susan Green Jones, "Ice Cream, Sugar and Slavery up to the 1900s."

100   Cralle invented the ice-cream scoop Samuel Momodu, "Alfred L. Cralle (1866–1920)," BlackPast.org, December 31, 2017, https://www.blackpast .org/african-american-history/cralle-alfred-l-1866-1920/.

101   social reformers also disdained Quinzio, *Of Sugar and Snow*, 104.

101   "From morning until night" Grace M. Mayer, *Once upon a City* (New York: Macmillan, 1958), 382.

102   "African article will not bear" Quinzio, *Of Sugar and Snow*, 104.

102   Most arrived without Quinzio, *Of Sugar and Snow*, 105.

103   When a study revealed Quinzio, *Of Sugar and Snow*, 109.

103 **In 1902, Harry Burt** Quinzio, *Of Sugar and Snow*, 174–75.

104 **image of the clean and wholesome** Quinzio, *Of Sugar and Snow*, 174–75.

104 **head of Columbia movie studios** Laura B. Weiss, *Ice Cream: A Global History* (London: Reaktion Books, 2011), 87.

104 **A stand at Coney Island** Quinzio, *Of Sugar and Snow*, 175.

104 **Epperson's company sold** Ralph Selitzer, *The Dairy Industry in America* (New York: Dairy and Ice Cream Field and Books for Industry, 1976), 266–67.

105 **"On a bright exhilarating day"** Mayer, *Once upon a City*, 395.

105 **"a drink that combines"** Weiss, *Ice Cream*, 63.

105 **over 475 million gallons** Weiss, *Ice Cream*, 63.

105 **popularity of soda fountains** Weiss, *Ice Cream*, 56.

105 **made use of its ice cellars** Weiss, *Ice Cream*, 57.

105 **By 1929, 60 percent** Quinzio, *Of Sugar and Snow*, 178.

106 **drugstore's revelatory "Cherry Sunday"** Michael Turback, *A Month of Sundaes* (New York: Red Rock Press, 2002), 30–32.

106 **"little girls and dudes drink ice cream"** Weiss, *Ice Cream*, 57.

106 **catered especially to female customers** Quinzio, *Of Sugar and Snow*, 124–25.

106 **"If he had served us ice cream"** "Royal Ice Cream Sit-In, 1957," Durham Civil Rights Heritage Project, September 18, 2020, https://durhamcounty library.org/exhibits/dcrhp/events/royal_ice_cream_sit_in_1957/.

107 **By the early 1970s** Quinzio, *Of Sugar and Snow*, 205.

108 **production of a single pint** "If It's Melted, It's Ruined," Ben & Jerry's, May 7, 2015, https://www.benjerry.com/values/issues-we-care-about/climate -justice.

108 **"Our world is already seeing"** "If It's Melted, It's Ruined," Ben & Jerry's.

108 **"could save up to 25 tonnes"** "Zero-Emission Refrigeration for Transporting Ice Creams Comes a Step Closer," Unilever, September 28, 2021, https://www.unilever.com/news/news-search/2021/zero-emission -refrigeration-for-transporting-ice-creams-comes-a-step-closer/.

109 **over 1.3 billion gallons** "Ice Cream Sales and Trends," International Dairy Foods Association, accessed October 17, 2022, https://www.idfa .org/ice-cream-sales-trends.

109 **"a symbol of the American dream"** Funderburg, *Chocolate, Strawberry, and Vanilla*, 161.

CHAPTER 6 · "CULTURE IN A GLASS"

113   **sweet tea "isn't a drink"** Allison Glock, "Sweet Tea: A Love Story," *Garden & Gun*, July/August 2008, https://gardenandgun.com/feature/sweet-tea-a-love-story.

113   **"a kind of culinary-cultural"** Quoted in John Shelton Reed and Dale Volberg Reed with William McKinney, *Holy Smoke: The Big Book of North Carolina Barbecue* (Chapel Hill: University of North Carolina Press, 2008), 192.

113   **"a rite of passage"** Colleen Thompson, "Elixir of the South," *Wrightsville Beach Magazine*, July 24, 2018, accessed October 17, 2022, https://wrightsvillebeachmagazine.com/elixir-of-the-south.

115   **"Tea made strong"** Quoted in Robert Moss, "Why 'As Southern as Sweet Tea' Isn't Very Southern at All," Serious Eats, June 15, 2020, https://www.seriouseats.com/sweet-tea-origin-story-history-south.

115   **"Our saloon keepers"** Moss, "Why 'As Southern as Sweet Tea' Isn't Very Southern at All."

116   **"Iced tea with a slice"** Moss, "Why 'As Southern as Sweet Tea' Isn't Very Southern at All."

116   **"Twenty years ago"** Moss, "Why 'As Southern as Sweet Tea' Isn't Very Southern at All."

117   **shipping of 50,000 tons** "Biographical Sketches—Capt. Charles Stedman," in *History of Bureau County*, ed. Henry C. Bradsby (Chicago: World Publishing, 1885), 660–61.

117   **"150 tons of"** Chris Naffziger, "The Ice Trade That Kept St. Louis Beer Flowing," *St. Louis Magazine*, August 22, 2018, accessed October 17, 2022, https://www.stlmag.com/history/the-ice-trade-that-kept-st-louis-beer-flowing.

118   **"refrigeration plant installed"** Isaac Taylor, *Taylor Report*, June 1905 University Archives, Department of Special Collection, Washington University Libraries.

118   **More than 25 million pounds** Taylor, *Taylor Report*, University Archives.

119   **"probably more appreciated"** Joe Gray Taylor, *Eating, Drinking, and Visiting in the South: An Informal History* (Baton Rouge: Louisiana State University Press, 2008), 128.

119   **To be clear** Moss, "Why 'As Southern as Sweet Tea' Isn't Very Southern at All."

121   **"the house wine of the South"** Thanks to Robert Moss for the reminder: Moss, "Why 'As Southern as Sweet Tea' Isn't Very Southern at All."

121 **"seven out of eight Southerners"** "Southerners Do Love Their Sweet Tea," *Greensboro News and Record*, July 26, 1993, https://greensboro.com /southerners-do-love-their-sweet-tea/article_859ebb3a-b04b-563e -983c-e257b3e7778d.html.

121 **"it's not Southern tea"** Moss, "Why 'As Southern as Sweet Tea' Isn't Very Southern at All."

121 **"appeal lies in"** Jeffrey Klineman, "I Wish I Lived in a Land of Lipton . . . ," *Slate*, August 8, 2007, accessed October 17, 2022, https://slate.com/human -interest/2007/08/what-makes-southern-sweet-tea-so-special.html.

122 **"losing some of its currency"** John Shelton Reed, *Surveying the South: Studies in Regional Sociology* (Columbia: University of Missouri Press, 1993), 52.

CHAPTER 7 · ON THE ROCKS

125 **"We're going to be the first bar"** Robert Simonson, "TriBeCa Bar Will Have Its Own Iceworks," *New York Times*, October 12, 2010, https://dinersjournal.blogs.nytimes.com/2010/10/12/tribeca-bar -will-have-its-own-iceworks/.

126 **which houses an ice machine** Simonson, "TriBeCa Bar Will Have Its Own Iceworks."

126 **"bandsaw, chainsaw, chisels, hammers"** Simonson, "TriBeCa Bar Will Have Its Own Iceworks."

126 **"eccentric reservation system"** Robert Simonson, "Sasha Petraske, 42, Dies; Bar Owner Restored Luster to Cocktail Culture," *New York Times*, August 21, 2015, https://www.nytimes.com/2015/08/22/nyregion /sasha-petraske-bar-owner-who-revived-luster-to-cocktail-culture -around-the-world-dies-at-42.html.

127 **first commercial mechanical-ice facility** Elizabeth M. Williams and Chris McMillian, *Lift Your Spirits: A Celebratory History of Cocktail Culture in New Orleans* (Baton Rouge: Louisiana State University Press, 2016), 44.

128 **inventor of the eponymous gin fizz** Williams and McMillian, *Lift Your Spirits*, 45.

129 **"She averred that if everyone"** Williams and McMillian, *Lift Your Spirits*, 47.

129 **"Ramos serves a gin fizz"** David Wondrich, *Imbibe! From Absinthe Cocktail to Whiskey Smash, a Salute in Stories and Drinks to "Professor" Jerry Thomas, Pioneer of the American Bar* (New York: Penguin, 2015), 138.

129 **By the time Mardi Gras came** "Shake, Rattle, and Roll: The History of the Ramos Gin Fizz," Gambino's Bakery, November 6, 2019, https://gambinos .com/cocktail-history/history-ramos-gin-fizz/.

130 **Shaking a cocktail brings** Dave Arnold, *Liquid Intelligence: The Art and Science of the Perfect Cocktail* (New York: W. W. Norton, 2014), 80–81.

130 **Larger ice cubes also reduce** Paul Adams, "The Science of Ice in Cocktails," *SevenFiftyDaily*, July 5, 2018, https://daily.sevenfifty.com /the-science-of-ice-in-cocktails/#:~:text=When%20a%20bartender %20chills%20a,of%20magnitude%20more%20cooling%20power.

130 **telling reporters how to make** Wondrich, *Imbibe!*, 140.

131 **"shake and shake and shake"** Wondrich, *Imbibe!*, 140.

132 **Pabst sold its cheese line** Maureen Ogle, *Ambitious Brew: The Story of American Beer* (New York: Harcourt, 2006) 184.

133 **"Who reads an American book?"** Wondrich, *Imbibe!*, 8.

134 **"the first in the city to concoct"** Wondrich, *Imbibe!*, 12.

134 **"was perhaps as famous"** Wondrich, *Imbibe!*, 13.

134 **"He is a gentleman"** Wondrich, *Imbibe!*, 14.

134 **Thomas perfected his cocktail** Wondrich, *Imbibe!*, 15–18.

135 **The adoption of syrups** Wondrich, *Imbibe!*, 63.

135 **Instead, he traveled** Wondrich, *Imbibe!*, 22.

137 **bartender Chris McLeod** Quoted in Paul Adams, "The Science of Ice in Cocktails," *SevenFiftyDaily*, July 5, 2018, https://daily.sevenfifty.com /the-science-of-ice-in-cocktails.

137 **So important is this process** Adams, "The Science of Ice in Cocktails."

137 **The Ice Doctor in Gainesville** Camper English, "The New Business of Big Ice," *SevenFiftyDaily*, August 8, 2017, https://daily.sevenfifty.com /the-new-business-of-big-ice/.

137 **Névé Ice in Los Angeles** Katie Robbins, "The Iceman Cometh: The Rise of the Gourmet Ice Entrepreneur," *Atlantic*, January 20, 2011, https://www.theatlantic.com/health/archive/2011/01/the-iceman -cometh-the-rise-of-a-gourmet-ice-entrepreneur/69804/.

137 **cornered its home cocktail market** English, "The New Business of Big Ice."

CHAPTER 8 · FIRE AND ICE

145 **The festival used** S.B., "At a Music Festival in Norway, the Instruments Are Made from Ice," *Economist*, February 11, 2020, accessed October 17, 2022, https://www.economist.com/prospero/2020/02/11/at-a -music-festival-in-norway-the-instruments-are-made-from-ice.

149 **The newlyweds would arrive** Ann Agranoff and Fred Anderes, *Ice Palaces* (New York: Abbeville Press, 1983), 7.

NOTES

149   The castles made their appearance Agranoff and Anderes, *Ice Palaces*, 23.

150   "quaint old market-place" Agranoff and Anderes, *Ice Palaces*, 23–24.

150   "were enough to make it gleam" Agranoff and Anderes, *Ice Palaces*, 27.

151   The city tried building Agranoff and Anderes, *Ice Palaces*, 34.

151   established the St. Paul Winter Carnival Agranoff and Anderes, *Ice Palaces*, 34.

152   colossal statue of King Borealis Agranoff and Anderes, *Ice Palaces*, 57.

153   The couple arrived escorted Agranoff and Anderes, *Ice Palaces*, 60.

153   That spectacle was shadowed Agranoff and Anderes, *Ice Palaces*, 66.

154   The Panic of 1893 Agranoff and Anderes, *Ice Palaces*, 79.

154   "People in the valleys" Agranoff and Anderes, *Ice Palaces*, 79–80.

154   Their highfalutin hopes Agranoff and Anderes, *Ice Palaces*, 79–80.

155   5,000 tons of ice Agranoff and Anderes, *Ice Palaces*, 80.

155   opened on New Year's Day Agranoff and Anderes, *Ice Palaces*, 81.

155   "Nothing ever appeared" Agranoff and Anderes, *Ice Palaces*, 81.

156   even that lower figure Agranoff and Anderes, *Ice Palaces*, 83.

158   "snow cleared from roads" Renee M. Lang et al., "Processing Snow for High Strength Roads and Runways," *Cold Regions Science and Technology* 25, no. 1 (January 1997): 167–74.

158   They built domes Agranoff and Anderes, *Ice Palaces,* 123.

CHAPTER 9 · FANCY FIGURES

164   fluidity by developing "grapevines" James R. Hines, *Figure Skating: A History* (Urbana: University of Illinois Press, 2006), 45.

164   "well-matched pair of horses" Hines, *Figure Skating*, 27.

165   Ice harvesters had to protect Hines, *Figure Skating*, 41.

165   "If it had been possible" Mariana Gosnell, *Ice: The Nature, the History, and the Uses of an Astonishing Substance* (New York: Alfred A. Knopf, 2005), 349.

165   "Ten years ago a lady" Howard Shubert, *Architecture on Ice: A History of the Hockey Arena* (Montreal: McGill-Queen's University Press, 2016), 22.

165   "What a charming thing" Shubert, *Architecture on Ice*, 22.

166   "The girls skate with one" Shubert, *Architecture on Ice*, 23.

166   "rich, warm materials" Shubert, *Architecture on Ice*, 43.

167   "The water next" Shubert, *Architecture on Ice*, 43.

167 **"No parlor etiquette"** Shubert, *Architecture on Ice*, 24.

168 **His fellow skaters found the sight** Hines, *Figure Skating*, 51–53.

169 **These performances were among** Erin Blakemore, "The Man Who Invented Figure Skating Was Laughed Out of America," History, August 22, 2018, https://www.history.com/news/the-man-who -invented-figure-skating-was-laughed-out-of-america.

169 **The style became known** Hines, *Figure Skating*, 51–53.

169 **"In remembrance of the American skating king"** Blakemore, "The Man Who Invented Figure Skating."

170 **"This little book"** Charlotte Oelschlägel, *Hippodrome Skating Book: Practical Illustrated Lessons in the Art of Figure Skating* (New York: The Hippodrome Skating Club, 1916), 3.

171 **"The ladies did the same"** Hines, *Figure Skating*, 92.

171 **"When are you guys"** Dvora Meyers, "Figure Skating Has Always Blurred the Lines of Gender Segregation," *FiveThirtyEight*, February 11, 2022, https://fivethirtyeight.com/features/in-a-sports-world-segregated -by-gender-figure-skating-has-often-blurred-those-lines.

172 **"A mingling of the two"** Shubert, *Architecture on Ice*, 47.

173 **coated with hog fat and soda** Shubert, *Architecture on Ice*, 78.

176 **"thousands of gas jets"** "February 12, 1879: New York Celebrates Its First Artificial Ice Rink at Madison Square Garden," Daily Dose, February 12, 2016, http://www.awb.com/dailydose/?p=1891.

176 **"We who have enjoyed"** "February 12, 1879: New York Celebrates," Daily Dose.

177 **find a skater to star** Roy Blakey, "Charlotte: Broadway's Skating Superstar," Icestage Archive, accessed December 26, 2022, https://www .icestagearchive.com/charlotte-broadways-skating-superstar/.

177 **first skater to appear in cinema** Hines, *Figure Skating*, 138.

179 **dreamed of an ice carnival** Hines, *Figure Skating*, 142.

179 **employed seventy-five performers** Bryan Curtis, "The Ice Capades: Requiem for the Ice Carnival," *Slate*, March 16, 2005, https://slate.com /news-and-politics/2005/03/the-life-and-death-of-the-ice-capades .html

179 **pets' vital statistics** Curtis, "The Ice Capades."

179 **Their "average height"** Curtis, "The Ice Capades."

179 **"Las Vegas revue on ice"** Curtis, "The Ice Capades."

179 **Producers did their best** Curtis, "The Ice Capades."

NOTES

CHAPTER 10 · CORK-BUNGS, BROOMS, AND ZAMBONIS

183 **The twenty-five-second shot** Lucas Anderson, "Thomas Edison Was the First to Film Hockey in 1898," The Hockey Writers, June 23, 2020, https://thehockeywriters.com/thomas-edison-first-hockey-film-1898/.

184 **"Concurrently with skating races"** J. M. Heathcote and C. G. Tebbutt, *Skating* (Boston: Little, Brown, 1892), 431.

185 **Other early stick-and-ball games** Howard Shubert, *Architecture on Ice: A History of the Hockey Arena* (Montreal: McGill-Queen's University Press, 2016), 48.

185 **"in good humour, health, and spirits"** "First Recorded Hockey Game in the Northwest Territories," Prince of Wales Northern Heritage Centre, October 18, 2022, https://www.pwnhc.ca/item/first-recorded-hockey -game-in-the-northwest-territories/.

186 **"accidents were likely to occur"** John Kalbfleisch, "Second Draft: In 1875, at the First Indoor Hockey Game, Guess What Broke Out?," *Montreal Gazette*, March 4, 2016, https://montrealgazette.com/opinion/columnists/second -draft-in-1875-at-the-first-indoor-hockey-game-guess-what-broke-out.

186 **"Shins and heads were battered"** Kalbfleisch, "Second Draft."

187 **eight images of twirling** Shubert, *Architecture on Ice*, 66.

188 **A report called** "1895–1896 Pittsburgh's Schenley Park Casino," PittsburghHockey.net, accessed August 22, 2022, https://pittsburghhockey.net.

188 **And in the *Pittsburgh Post*** "1895–1896 Pittsburgh's Schenley Park Casino," PittsburghHockey.net.

188 **"didn't know just what"** "1885–1896 Pittsburgh's Schenley Park Casino," PittsburghHockey.net.

188 **It caused an explosion** "1895–1896 Pittsburgh's Schenley Park Casino," PittsburghHockey.net.

189 **eliminated the uncertainty of weather** Shubert, *Architecture on Ice*, 85.

189 **"women spectators in evening gowns"** Shubert, *Architecture on Ice*, 88.

190 **Four employees would follow** "The Zamboni Story," Zamboni, https://zamboni.com/about/zamboni-archives/the-zamboni-story/2/.

190 **built on a Jeep frame** Cait Murphy, "This Is How the First Zamboni Machine Was Invented," *Time*, October 11, 2016, https://time.com /4508517/zamboni-history/.

191 **When the machine appeared** Murphy, "This Is How the First Zamboni Machine Was Invented."

191 **"There are three"** "Snoopy® and Zamboni," Zamboni.com, https:// zamboni.com/news-media/media-mentions/snoopy.

191   **it made the ice safer** Mariana Gosnell, *Ice: The Nature, the History, and the Uses of an Astonishing Substance* (New York: Alfred A. Knopf, 2005), 348.

191   **hockey ice should be** "Miscellaneous Trivia," National Hockey League, accessed August 19, 2022, http://www.nhl.com/ice/page.htm.

191   **not unusual to find** Gosnell, *Ice*, 347.

193   **The artwork depicts** Gosnell, *Ice*, 355.

195   **In a popular YouTube series** "The Controversial Physics of Curling: Cold Hard Science," Smarter Every Day, March 13, 2014, https://www.youtube .com/watch?v=7CUojMQgDpM.

CHAPTER 11 · THE NEED FOR SPEED

201   **"Reality is slippery"** Mariana Gosnell, *Ice: The Nature, the History, and the Uses of an Astonishing Substance* (New York: Alfred A. Knopf, 2005), 358.

201   **that ice is slippery because** Frank Philip Bowden and T. P. Hughes, "The Mechanism of Slide on Ice and Snow," *Proceedings of the Royal Society* 172, no. 949 (August 3, 1939): 280–98, https://doi.org/10.1098 /rspa.1939.0104.

202   **the skater essentially hydroplanes** Bob Rosenberg, "Why Is Ice Slippery?," *Physics Today* 58, no. 12 (December 2005): 50, https://doi.org/10.1063 /1.2169444.

202   **"Different experiments often have"** Yuen Yiu, "Why Is Ice Slippery? It's Not a Simple Question," *Inside Science*, May 20, 2021, https://www.insidescience.org/news/why-ice-slippery-its-not-simple -question.

204   **"It has been said"** Reed Sparling, "Newburgh: The Cradle of American Speedskating," Scenic Hudson, accessed December 26, 2022, https://www .scenichudson.org/viewfinder/newburgh-americas-speedskating-capital/.

205   **"a strong wind"** Tracy Ziemer, "Long Blades, Fast Times: How Newburgh Changed Speedskating," *Times Union*, February 12, 2022, https://www .timesunion.com/hudsonvalley/outdoors/article/Newburgh-epicenter -speedskating-16847241.php.

205   **"with ridiculous ease"** Stephan Benzkofer, "When Speed Skating Was King of Winter," *Chicago Tribune*, February 16, 2014, https://www .chicagotribune.com/news/ct-ice-skating-chicago-flashback-0216 -20140216-story.html.

206   **Silver Skates competition grew** Benzkofer, "When Speed Skating Was King of Winter."

CHAPTER 12 · FEVERS, FREEZERS, AND FRANKENSTEIN

215 **vaccines were deemed unusable** "Pfizer-BioNTech COVID-19 Vaccine: Thermal Shipping Container Dry Ice Replenishment Instructions," Ohio Department of Health, last updated December 12, 2020, https://odh.ohio .gov/static/covid19/vaccine-providers/replenishing-dry-ice-for-pfizer -biontech-covid-19-vaccine.pdf.

215 **estimated that nearly half a million** "WHO/ECDC: Nearly Half a Million Lives Saved by COVID-19 Vaccination in Less Than a Year," European Centre for Disease Prevention and Control, November 25, 2021, https://www.ecdc.europa.eu/en/news-events/who-ecdc-nearly-half -million-lives-saved-covid-19-vaccination.

217 **American attitudes toward cold** Robert F. Reilly, "Medical and Surgical Care during the American Civil War, 1861–1865," *Baylor University Medical Center Proceedings* 29, no. 2 (April 2016): 138–42, doi:10.1080 /08998280.2016.11929390.

218 **Three out of every four** Reilly, "Medical and Surgical Care during the American Civil War."

218 **lead to the transmission** Reilly, "Medical and Surgical Care during the American Civil War."

219 **considered ice a necessity** Terry Reimer, "Medical Uses of Ice," National Museum of Civil War Medicine, November 9, 2003, https://www .civilwarmed.org/surgeons-call/ice/.

219 **"seized with dimness"** George Alexander Otis, Joseph Janvier Woodward, Charles Smart, David Lowe Huntington, and Joseph K. Barnes, *The Medical and Surgical History of the War of the Rebellion (1861–65)* (Washington, D.C.: Government Printing Office, 1870), 425.

220 **"Should the ice prove"** Otis et al., *The Medical and Surgical History of the War of the Rebellion*, 425.

220 **"melted in the cars"** Otis et al., *The Medical and Surgical History of the War of the Rebellion*, 425.

221 **extreme cold had reduced** Michael A. Bohl, Nikolay L. Martirosyan, Zachary W. Killeen, Evgenii Belykh, Joseph M. Zabramski, Robert F. Spetzler, and Mark C. Preul, "The History of Therapeutic Hypothermia and Its Use in Neurosurgery," *Journal of Neurosurgery* 130, no. 3 (March 2019): 1014, https://doi.org/10.3171/2017.10.JNS 171282.

222 **their chance of survival** Bohl et al., "The History of Therapeutic Hypothermia and Its Use in Neurosurgery," 1014.

223 **"The wide application"** Bohl et al., "The History of Therapeutic Hypothermia and Its Use in Neurosurgery," 1016.

224 **"was not a happy monkey"** Phil Jaekl, *Out Cold: A Chilling Descent into the Macabre, Controversial, Lifesaving History of Hypothermia* (New York: PublicAffairs, 2021), 194.

225 **therapeutic hypothermia is practiced** Jaekl, *Out Cold*, 210.

225 **"He was almost dead"** "Man 'Frozen' to Save His Life," ABC News, March 31, 2005, accessed August 15, 2022, https://abcnews.go.com/GMA/Health/story?id=628745.

226 **He spent weeks** Jaekl, *Out Cold*, 210.

226 **"controlled destruction of tissue"** S. M. Cooper and R.P.R. Dawber, "The History of Cryosurgery," *Journal of the Royal Society of Medicine* 94, no. 4 (April 2001): 196, https://www.ncbi.nlm.nih.gov/pmc/articles/PMC1281398.

231 **"At a time when surgical excision"** Cooper and Dawber, "The History of Cryosurgery," 196.

CHAPTER 13 · TV DINNERS, HENRY DAVID THOREAU, AND A SWIFTLY WARMING PLANET

235 **"10,000 tons of ice"** Stephen Hahn, "Commonplace Commerce and Transcendence in 'The Pond in Winter,'" Thoreau Society Annual Gathering, 2016, https://commons.digitalthoreau.org/tsag2016/thursday-july-7-2/commonplace-commerce-and-transcendence-in-the-pond-in-winter/.

236 **"Climate change has come"** Richard B. Primack, *Walden Warming: Climate Change Comes to Thoreau's Woods* (Chicago: University of Chicago Press, 2014), ix.

236 **Refrigerators and stand-alone freezers** Jeffrey Kluger, "How the Ice in Your Drink Is Imperiling the Planet," *Time*, April 14, 2011, https://science.time.com/2011/04/14/how-the-ice-in-your-drink-is-imperiling-the-planet.

236 **If the fridge includes** Jeffrey Kluger, "How the Ice in Your Drink Is Imperiling the Planet."

236 **That's about half** Rick Adams, "Your Fridge Could Be Impacting Global Warming," Spectrum News 1, November 25, 2021, https://spectrumnews1.com/ca/la-west/environment/2021/11/25/your-fridge—-global-warming-.

236 **one of the fastest lifestyle** Gina Medsker, "Refrigerators: 1940s–1950s," Evolution of Home Appliances in the US, accessed August 15, 2022, https://evolutionhomeappliances.weebly.com/refrigerators-1940s-1950s-new-features-and-the-deep-freeze.html.

236 **only 9 percent** "Number of TV Sets in America: 1950–1978," The American Century, accessed August 18, 2022, https://americancentury.omeka.wlu.edu/items/show/111.

NOTES

237 **came to signal** "Number of TV Sets in America: 1950–1978," The American Century.

237 **By the mid-1930s** Jonathan Rees, *Refrigeration Nation: A History of Ice, Appliances, and Enterprise in America* (Baltimore: Johns Hopkins University Press, 2013), 137.

237 **"We know of no cheap way"** Rees, *Refrigeration Nation*, 137.

237 **Commercial refrigerators were also** Rees, *Refrigeration Nation*, 137.

238 **"We already have a number"** Rees, *Refrigeration Nation*, 138.

238 **Between 1914 and 1922** "DOMELRE First Electric Refrigerator," ASHRAE, accessed August 18, 2022, https://www.ashrae.org/about /mission-and-vision/ashrae-industry-history/domelre-first-electric -refrigerator.

239 **"A little water is put"** *New Yorker*, quoted in Sylvia Lovegren, *Fashionable Food: Seven Decades of Food Fads* (New York: Macmillan, 1995), 10.

239 **"the Magi and the Christ child"** Rees, *Refrigeration Nation*, 142.

239 **The volume of gas** Rees, *Refrigeration Nation*, 145.

239 **"servicemen acquired strange reputations"** Rees, *Refrigeration Nation*, 145.

240 **average family doubled** Rees, *Refrigeration Nation*, 145.

240 **sold with recipe books** Rees, *Refrigeration Nation*, 156.

240 **government loans to purchase** Rees, *Refrigeration Nation*, 165.

240 **only 30 percent** Rees, *Refrigeration Nation*, 162.

240 **the last icebox manufacturer** Rees, *Refrigeration Nation*, 137.

241 **food shortages that forced** Rees, *Refrigeration Nation*, 173.

241 **families were making trips** Rees, *Refrigeration Nation*, 173.

241 **"roomy interior"** Rees, *Refrigeration Nation*, 172.

242 **one in four Americans owned** "What's New in How We Use Energy at Home: Results from EIA's 2015 Residential Energy Consumption Survey (RECS)," U.S. Energy Information Administration, May 2018, accessed August 18, 2022, https://www.eia.gov/consumption/residential/reports /2015/overview/pdf/whatsnew_home_energy_use.pdf.

242 **Freezer aisles are in** Heather Long, "23% of American Homes Have Two (or More) Fridges," CNN, May 27, 2016, https://money.cnn.com/2016/05/27 /news/economy/23-percent-of-american-homes-have-2-fridges/index.html.

242 **Like the ice trade** David B. Goldstein, "Refrigerator Reform: Guidelines for Energy Gluttons," *Technology Review* 86, no. 2 (Massachusetts Institute of Technology), February–March 1983, 36.

242 **two hundred countries signed** "About Montreal Protocol," United Nations Environment Programme, accessed August 18, 2022, https://www.unep .org/ozonaction/who-we-are/about-montreal-protocol.

243 **occur at the end** "Refrigerant Management," Project Drawdown, accessed August 18, 2022, https://drawdown.org/solutions/refrigerant -management.

243 **models that didn't conform** Goldstein, "Refrigerator Reform," 38.

243 **set of proposed standards** Goldstein, "Refrigerator Reform," 41

243 **elimination of standards** Goldstein, "Refrigerator Reform," 41.

244 **"limits the freedom"** "Memorandum of Disapproval of the Appliance Energy Conservation Bill," November 1, 1986, Ronald Reagan Presidential Library, https://www.reaganlibrary.gov/archives/speech/memorandum -disapproval-appliance-energy-conservation-bill.

244 **"ranks last among"** Goldstein, "Refrigerator Reform," 43.

244 **"Manufacturers today face"** Goldstein, "Refrigerator Reform," 46.

245 **"carbon bomb"** Phil McKenna, "I Tried to Buy a Climate-Friendly Refrigerator. What I Got Was a Carbon Bomb," *Inside Climate News*, March 11, 2021, accessed August 18, 2022, https://insideclimatenews .org/news/11032021/climate-change-refrigerator-hfc-super-pollutant/.

246 **Some of these new standards** Joseph A. Davis, "Biden Rollbacks of Trump Energy-Efficiency Cuts Generate Local Stories," Society of Environmental Journalists, March 3, 2021, accessed August 18, 2022, https://www.sej.org /publications/tipsheet/biden-rollbacks-trump-energy-efficiency-cuts -generate-local-stories.

246 **national coalition of utilities** John W. Feist, Ray Farhang, Janis Erickson, Elias Stergakos, Paul Brodie, and Paul Liepe, "Super Efficient Refrigerators: The Golden Carrot from Concept to Reality," American Council for an Energy-Efficient Economy, Summer 1994, https://www .aceee.org/files/proceedings/1994/data/papers/SS94_Panel3 _Paper08.pdf.

247 **most energy efficient fridges** John Holusha, "Whirlpool Takes Top Prize in Redesigning Refrigerator," *New York Times*, June 30, 1993, https://www .nytimes.com/1993/06/30/business/company-news-whirlpool-takes-top -prize-in-redesigning-refrigerator.html.

247 **The United States Senate** Robinson Meyer, "The Senate Just Quietly Passed a Major Climate Treaty," *Atlantic*, September 28, 2022, https:// www.theatlantic.com/science/archive/2022/09/congress-climate-policy -hydrofluorocarbons-kigali-amendment/671579.

249 **Another potentially game-changing technology** P. Lloveras, A. Aznar, M. Barrio, Ph. Negrier, C. Popescu, A. Planes, L. Mañosa, et al., "Colossal Barocaloric Effects Near Room Temperature in Plastic Crystals of

Neopentylglycol," *Nature Communications* 10, no. 1803 (April 2019), https://www.nature.com/articles/s41467-019-09730-9.

249 **"Maybe in the future"** Heidi Julavits, "American Exceptionalism on Ice," *New Yorker*, July 8, 2016, https://www.newyorker.com/culture/culture-desk/american-exceptionalism-on-ice.

CHAPTER 14 · FLAMMABLE ICE

253 **nearly 8 million tons** Laura Parker, "The World's Plastic Pollution Crisis Explained," *National Geographic*, updated May 20, 2022.

254 **"60 percent of Barrow's"** Eli Kintisch, "These Ice Cellars Fed Arctic People for Generations. Now They're Melting," *National Geographic*, October 30, 2015.

254 **average air temperatures** Kintisch, "These Ice Cellars Fed Arctic People for Generations."

254 **As the permafrost melts** Kintisch, "These Ice Cellars Fed Arctic People for Generations."

254 **The Iñupiat are adapting** Kintisch, "These Ice Cellars Fed Arctic People for Generations."

254 **the traditional knowledge** Kintisch, "These Ice Cellars Fed Arctic People for Generations."

255 **This "savior" ice** Richard Anderson, "Methane Hydrate: Dirty or Energy Saviour?," BBC, April 17, 2014, https://www.bbc.com/news/business-27021610.

256 **Hydrates also release less** Martha Henriques, "Why 'Flammable Ice' Could Be the Future of Energy," BBC, November 22, 2018, https://www.bbc.com/future/article/20181119-why-flammable-ice-could-be-the-future-of-energy.

256 **"provide an answer"** Anderson, "Methane Hydrate."

257 **A second test completed** Henriques, "Why 'Flammable Ice' Could Be the Future of Energy."

257 **"climate time bombs"** Todd Woody, "Huge Amounts of Greenhouse Gases Lurk in the Oceans, and Could Make Warming Far Worse," *National Geographic*, December 17, 2019, https://www.nationalgeographic.com/science/article/greenhouse-gases-lurk-in-oceans-could-make-warming-far-worse.

257 **She explained this idea** Carolyn D. Ruppel and John D. Kessler, "The Interaction of Climate Change and Methane Hydrates," *Reviews of Geophysics* 55, no. 1 (March 2017): 138, https://doi.org/10.1002/2016RG000534.

CONCLUSION

264 **Today, most of those** Scottie Andrew, "Some of Glacier National Park's Glaciers Have Lost as Much as 80% of Their Size in the Last 50 Years," CNN, September 16, 2020, https://www.cnn.com/travel/article/glacier -national-park-melting-scn-trnd/index.html.

264 **In Switzerland, glaciers** Jack Guy, "Record Melting Sees Swiss Glaciers Shrink 10% in Five Years," CNN, October 15, 2019, https://www.cnn.com /2019/10/15/europe/switzerland-glacier-melting-scli-intl-scn/index.html.

264 **The Himalayas are often** Emma Newburger, "Himalayan Glaciers Are Melting at an Extraordinary Rate, Research Finds," CNBC, December 20, 2021, https://www.cnbc.com/2021/12/20/himalayan-glaciers-melting -at-extraordinary-rate-research-finds-.html.

264 **In 2019, the Icelandic** Harmeet Kaur, "Scientists Bid Farewell to the First Icelandic Glacier Lost to Climate Change. If More Melt, It Can Be Disastrous," CNN, August 18, 2019, https://www.cnn.com/2019/08/18 /health/glaciers-melting-climate-change-trnd/index.html.

264 **"This monument is to"** Jon Henley, "Icelandic Memorial Warns Future: 'Only You Know If We Saved Glaciers,'" *Guardian*, July 22, 2019, https:// www.theguardian.com/environment/2019/jul/22/memorial-to -mark-icelandic-glacier-lost-to-climate-crisis.

265 **The signs now read** Christina Maxouris and Andy Rose, "Glacier National Park Is Replacing Signs That Predicted Its Glaciers Would Be Gone by 2020," CNN, January 8, 2020, https://www.cnn.com/2020/01/08/us /glaciers-national-park-2020-trnd/index.html.

PHOTO CREDITS

TITLE PAGE · *Page ii*

Two young women deliver ice during World War I (1918):
National Archives, photo no. 165-WW-595A-3

INTRODUCTION · *Page xi*

Harold "Red" Grange, All-American halfback, delivers ice during the off-season (1930):
Associated Press

CHAPTER 1 · *Page 3*

Two men harvesting ice in Michigan (1903):
Prints & Photographs Division, Library of Congress, LC-DIG-det-4a05655

CHAPTER 2 · *Page 31*

Ice harvesting on Conneaut Lake, Pennsylvania (1907):
Prints & Photographs Division, Library of Congress, LC-USZ62-71467

CHAPTER 3 · *Page 51*

Handsomest Iceman contestant being powdered by a woman (1940):
New York World's Fair 1939–1940 records, Manuscripts and Archives Division, New York Public Library

CHAPTER 4 · *Page 71*

People gathered around pies on a Coca-Cola cooler during the wheat harvest in Nebraska (1959):
Warren K. Leffler/Prints & Photographs Division, Library of Congress, LC-U9-2698- 20

CHAPTER 5 · *Page 89*

Cuban ice-cream peddlers (c. 1900):
ilbusca/DigitalVision Vectors via Getty Images

PHOTO CREDITS

CHAPTER 6 · *Page 111*

An iceman waits for a phone call for delivery (1915):
The Protected Art Archive/Alamy Stock Photo

CHAPTER 7 · *Page 123*

A bartender serving another drink:
The Protected Art Archive/Alamy Stock Photo

CHAPTER 8 · *Page 141*

The St. Paul Winter Carnival Ice Palace (1887):
Minnesota Historical Society

CHAPTER 9 · *Page 161*

Norwegian skating champion Sonja Henie training at the Palais des Sports
in Paris (1936):
Associated Press

CHAPTER 10 · *Page 181*

A hockey player on a frozen lake (1940s):
H. Armstrong Roberts/ClassicStock via Getty Images

CHAPTER 11 · *Page 199*

Larry Jensen competes in the juvenile boys' quarterfinal event during the
Silver Skates Derbies (1954):
Chicago Tribune/TCA

CHAPTER 12 · *Page 213*

Ad for a J. C. Taylor & Son's "Cold Air Ice Casket" (1879):
The Casket *magazine*

CHAPTER 13 · *Page 233*

Mrs. Pleas Rodden puts fresh milk in her Frigidaire in West Carroll Parish,
Louisiana (1940):
*Marion Post Wolcott/Prints & Photographs Division, Library of Congress,
LC-USF34-053893-E*

CHAPTER 14 · *Page 251*

A climate change protest in Union Square, San Francisco (2019):
Li-An Lim via Unsplash

INDEX

Note: Page numbers in *italics* indicate illustrations.

A

Adams, Mary Louise, 171
"Admiral of the Atlantic," 66
air conditioners, 41
"All She Gets from the Iceman Is Ice,"
    53–54
Alpine Room, 173
*Ambitious Brew: The Story of American
    Beer* (Ogle), 132
American Ice Company, 64–66
Anheuser-Busch, 26, 105, 116–117, 132
Anniston Star, 121
Apalachicola, Florida, 34–35, 38
Arkansas Democratic-Gazette, 121
Arnold, Dave, 129–130
artificial ice. *See* manufactured ice
*Artistic Impressions: Figure Skating,
    Masculinity, and the Limits of Sport*
    (Adams), 171
asymmetrical friction, 195
Avicenna, 216

B

bandy, 184
"Barefoot Blue Jean Night" (Owen), 114
Barnum, Phineas T., 44, 174–175
Basile, Salvatore, 39
Baskin-Robbins, 107

Beagle, Peter S., 80
beer and breweries
    Anheuser-Busch, 26, 105, 116–117, 132
    Coors, 132
    German family lager recipes, 25
    Pabst Brewing Company, 26, 132
    popularity of, 25
    during Prohibition, 132
    St. Louis and, 116–119
Bellazzi, Franz, 168–169
Bell in Hand Tavern, 5
Beman, Caroline Frances Myrick, 42
Ben & Jerry's, 108
Berkeley, Norborne, 93
Berry, Chuck, 80
Bishop, Lou, 154
Black, William, 94
"the black vomit." *See* yellow fever
Blechynden, Richard, 119
"block busters," derivation of term, 49
block ice, 84
Boccanfuso, Alexis, 197
Boccato, Richard, 125–126
*Bombay Times and Journal of
    Commerce,* 43
Book of Ice-Cream (Fisk), 107
Boston Gazette, 16
Boston Journal, 116
Bowden, Frank Philip, 201

Brain Research Laboratory (BRL), 223–224
Branch, John, 190–191
Brenston, Jackie, 80
British icehouses, xvii
Brown, William Alexander, 98–99
Brown, Zac, 113–114
Bruegel, Pieter, the Elder, 193
Bryan, Lettice, 114
Burt, Harry, 103–104
Bushnell, Edward W., 204

C

café culture, 19
Cambridge Enterprise, 249
Canada, 186
Canon of Medicine (Avicenna), 216
carnivals on ice, 172
Carroll, John, 65
cars
  ice docks and, 74
  road culture of, 80–82
  7-Eleven stores and, 79–80
Carson, Jack, 104
Casino (Pittsburgh), 187–188
Castle, Sheri, 113
Castro, Fidel, 20–21
Cavanagh, Ralph, 246–247, 248
Chadwick, Henry, 165
"chamiare," 184
Chapman, Alvan W., 42
Charles, Ray, 70
Charlotte Stop, 177
Chatham Artillery Punch, 114
Chicago, 205–206
Chicago Chronicle, 59
Chicago Daily Inter Ocean, 116
Chicago Tribune, 106, 205
Chicago World's Fair, 119–120
"Chicken Fried" (Zac Brown Band), 113–114
Chilly Billy, 78–79
China, ancient, xvii
chlorofluorocarbons (CFCs), 242, 247

Citizen Kane (movie), 144
citrus fruit and manufactured ice, 48
Civil War
  casualties, 33, 217–218
  manufactured ice and, 47
  medical care during, 217–220
  northern ice trade and southern hospitals, 33–34
climate change. *See also* environment
  glaciers and, 264, 265
  natural ice and, 236
  permafrost melting, 254
  public demonstrations, *251*
Clinebell ice-making machines, 126, 137, 138, 146
CNC (computer numerical control) machines, 146–147
Cockburn, Lord Henry, 164
cocktails
  gin fizz, 128–131
  importance of ice, 126
  before introduction of ice, 133
  invention of, 24
  during San Francisco gold rush, 135
"Cold Air Ice Caskets," *213*
Collot, Monsieur, 99
Commercial Advertiser, 43
confectionery shops, 96–97, 99–100, 105
Consolidated Ice Company, 61, 63
convenience stores, 79–80
coolers, *71*
  invention of, 81
  magnetic wine, 248
  road trips and, 73, 81
Cooper, S. M., 230–231
Coors (brewery), 132
cork-bungs, 185
corpses, preservation of, xvi
Covitz, Bill, 145–149
"the cradle of drinking culture." *See* New Orleans
"cream of almonde," 94
crema della mia nonna ("my grandmother's cream"), 94
Croker, Richard, 65

"crowd poison," 39, 40
Crowninshield, Mary Boardman, 95
Crudele, Caroline, 205
cryosurgery (cryotherapy, cryogenic
    surgery, cryoablation), 222, 226–231
Crystal Carnival Association, 154
Cuba
    Frederic and John Henry Tudor in, 6–8
    Frederic Tudor's marketing of ice in,
        19–20
    Frederic Tudor's success in, 19–20
    ice cream in, 20–21, *89*
    permissions to sell ice in, 10–11
    rum from, 133
    sale of ice in, 18
curling, 192–197
Currie, James, 216
Curry, Jabez Lamar Monroe, 34

D

Daily News, 165
Daily Witness, 186
Dairy Queen, 80
Daugherty, Harry, 68
Dawber, R. P. R., 230–231
Dayton, Abram C., 98
Delaney, Philip, 81
deliveries, *ii,* 55–56. *See also* icemen
De Marsan, H., 167
The Dentist (film), 54
Detroit River, 58
Detroit Tribune, 58
Dickens, Charles, xvi
Dillingham, Charles, 176–177
"directional freezing" of ice, 137–139
diseases, 39–40. *See also* yellow fever
"Doin' What She Likes" (Shelton), 114
DOMELRE (domestic electric
    refrigerator), 238
Donaldson, Gary, 84–85
Donoghue, Timothy Sr., 204–205
dry ice, 215
Dubois, Gwendolyn, 205
Dull, Mrs. S. R., 120

Durr, Nicholas, 227
Dutch Kills, 137

E

*Eating, Drinking, and Visiting in the South:
    An Informal History* (Taylor), 119
Edge, John T., 113
Edinburgh Review, 133
Edward Scissorhands (movie), 145
Egerton, John, 121
Egypt, ancient, 216
Emancipation Proclamation, 33
Emerson, Ralph Waldo, 235
energy
    of hydrates, 255
    refrigerators' use of, 236–237, 239–240,
        242, 244–248
Energy Star ratings, 245–246
England, medieval, 94
English, Camper, 138–139
environment. *See also* climate change
    impact on, of making ice cream, 108
    Kigali Amendment, 247
    plastic waste, 253
    Reagan and, 243–245
    refrigerators and, 236–237, 239–240,
        242–243
    water pollution, 57–59
Epperson, Frank, 104

F

Fancy Ice Carving (Forster), 145
Faraday, Michael, 202
fast food chains, 79, 122
Favorite, 15, 16
Fay, Temple, 220–223
feet, hypersexualization of, 166–167
"fever season." *See* yellow fever
Fields, W. C., 54
figure skating
    in Americas, 163–165
    clubs, 167
    in Europe, 163

figure skating (*cont.*)
  Haines and, 168–169
  Harris and, 178–179
  Henie, *161,* 178
  Ice Capades, 179–180
  at Olympics, 167, 171, 172
  Viennese/International style, 169–170
  by women, 165–167, 170, 177–178
  World Figure Skating Hall of Fame, 169
Figure Skating in the Formative Years:
  Singles, Pairs, and the Expanding Role
  of Women (Hines), 171
First Nations communities, 185. *See also*
  Indigenous communities
Fisk, Walter W., 107
"flammable ice," 255–259
Fleming, Peggy, 171
Flower, Roswell P., 62
Food and Drug Association (FDA), 83
food poisoning and iceboxes, 25
food preservation, xvii, 25
Forster, August, 145
Francisco (Tudor servant), 7–8
Franklin, Sir John, 185
freezer warehouses, 240–241
friction, asymmetrical, 195
Frigidaire, 76
Frozen (film), xv
fruit trade and manufactured ice, 48
Funderburg, Anne Cooper, 109

G

Gamgee, John, 173
García Márquez, Gabriel, 20
Garden & Gun (Glock), 113
gas station–convenience store combos, 76
gender. *See* women
General Electric (GE), 76–77, 239–240, 248
General Motors, 76
General Tom Thumb, 44
Gilmore, Patrick, 175
Gilmore's Garden, 176
gin fizz, 128–131
The Girl Can't Help It (film), 55
Glaciarium, 173

Glacier National Park, 264, 265
glaciers, 264, 265
Glock, Allison, 113
Goldstein, David B., 244
Good Humor Ice Cream Company, 103–104
The Good Humor Man (movie), 104
Gorman, T. P., 189
Gorrie, John
  application of ice to patients' bodies by,
    40–41
  basic facts about, 35–36, 42, 49–50
  "crowd poison" as cause of yellow
    fever, 40
  death of, 46–47
  experiments with manufacturing ice,
    41, 42
  invention of air conditioner, 41
  investments in manufacturing venture,
    44, 46
  manufactured ice and, 34
  in medical school, 37
  nom de plume of, 43
  temperature of human body and, 40
  treatment of yellow fever by, 37, 38
Grand Cathedral of Socialist Ice Cream,
  20–21
Grange, Harold "Red," *xi,* 55
Grant, Ulysses, xvi
Green, John "Uncle Johnny," 75
Green, Samuel, 36, 37
Green, Susan, 100
greenhouse gases, 236
Greenland ice cap, nuclear-powered Army
  camp on, 158
Grelen, Jay, 121

H

Häagen-Dazs, 107
Haier, 248
Haines, Jackson, 168–169
Hamilton, Scott, 171, 172
Handsomest Iceman contests, *51,* 78
The Handwriting on the Wall: A Call to
  Arms! (Natural Ice Association of
  America), 60

Harding, Warren G., 68

Harper's Weekly, 105

Harris, John H., 178–179

Harrisburg Telegraph, 116

Harrison, William Henry, 44

harvesting, *3, 31*

  banned in vicinity of Schuylkill River,
    57–58

  dangers of, xv, 10

  locations and health hazards of natural
    ice, xv, 57–59

  at Mount Vernon, 91–92

  Tudor's first, 15–16

health and safety

  avoidance of cold for, 217

  Good Humor man image, 104

  ice-cream peddlers and, 101, 102–103

  of natural ice, xv, 57–59

  of packaged ice, 82

  standards for ice, 82–84

  use of ice for, 216–217

Hearst, William Randolph, 63

Heinze, F. Augustus, 66–67

el helado, 20–21, *89*

Henie, Sonja, *161,* 178, 190

Hicks, John S., 100

Hines, James R., 171

Hip-Hip Hooray (show), 177

Hippodrome, 174–175

Hippodrome Theatre, 176–177

Holiday Inn, 82

Holland House, 131–132

Holmes, Miss, 99–100

*Holy Smoke: The Big Book of North Carolina
    Barbecue* (Edge), 113

Hopkinson, Elizabeth, 27–28

Horse Feathers (film), 55

horses, 55, 56

Hoshizaki machines, 136

Hu, Clarisse, 227

Hudson River, ice harvesting in
    vicinity of, 58

Hughes, T. P., 201

Humphreys, Mary Gay, 105

Hunters in the Snow (Pieter Bruegel the
    Elder), 193

Hutchison, Alexander Cowper, 150, 151

hydrofluorocarbons (HFCs), 242–243, 247

I

ice

  availability of, 265

  impurities in, 138, 207

  mindset of plentifulness of, 254

  slipperiness of, 195, 201–203, 207–208

  vaccine storage, 215

Ice and Refrigeration, 118

iceboxes

  described, 25

  DOMELRE and, 238

  importance of, 25

  invention of, 24

  refrigerators and, 76–77, 237, 240

  in restaurants and cafés, 120

Ice Capades (show), 179–180

ice castles. *See* ice palaces

ice cellars, 253–254

ice cream

  Ben & Jerry's, 108

  confectionery shops, 96–97, 99–100, 105

  cookbook recipes for, 95

  in Cuba, 20–21, *89*

  cups for eating, 92, 93

  Good Humor trucks, 103–104

  as luxury for wealthy, 92

  manufacture of, 107, 108–109

  for masses, 98–102

  origin of, 93–94

  peddlers of, *89,* 100–103

  pleasure gardens, 97–100

  popularization of, 94–96

  regulation of sale of, 103

  soda fountains, 105–106

Ice Cream: A Global History (Weiss), 105

ice dancing, 168–169

ice docks, 74, 75

Ice Doctor, 137

iced tea, 113–116, 118–120

ice hockey, *181*

  condition of ice and, 189–191

  early versions of, 184–185

ice hockey (*cont.*)
 Edison movie, 183
 ice rinks and, 186–188
 violence during, 186, 187, 188–189
Ice House, 84–85
icehouses
 British, xvii
 described, 9
 on Martinique, 16, 17
 at Monticello, xvi
 in New Orleans, 127
 storage system in, 10
ice industry, consolidation of,
  61–65
ice industry war, 26–27
"Ice King." *See* Tudor, Frederic
ice machines, 82
The Iceman Cometh (O'Neill), 54
Ice Manufacturing Co., 66
Ice Matters, 145
icemen, *xi, 111*
 contests for handsomest, *51,* 78
 in popular culture, 53–55
 refrigerators and, 78
 as symbols of male virility, 43–55,
  69–70
 tracking of inventory by, 56–57
 during World War I, 69
Ice Music Festival (Norway), 145
"ice obsessed," Americans identifying
  as, xiii
ice palaces
 Leadville, Colorado, 153–156
 Montreal, 149–151
 Russia, 149
 St. Paul Winter Carnival, *141,* 151–153,
  156–157
ice rinks
 building of permanent structures,
  172–173
 Dillingham and, 176–177
 Gilmore and, 175–176
 ice hockey on, 186–188
 manufactured ice for, 49, 173,
  175–176
 traveling carnivals on, 172

ice sculpting
 basic facts about, 143–144, 149
 CNC (computer numerical control)
  machines, 146–147
 wealth and, 144–145
Ice Sculpture Christmas (movie), 145
ice skating. *See also* figure skating; ice rinks
 as democratic sport, 167
 European history of, 163
 Iroquois and, 163
 speed skating, *199,* 203–206
ice tools, development of, 24
"the Ice Trust," 61, 64–66
ice wagons, 56
I Love Lucy (television program), 241
Imperial Cabinet, 128, 129
Indigenous communities
 early ice hockey and, 185
 importance of ice to Inuit, xvii
 Iñupiat ice cellars, xvii, 253–254
 Iroquois and ice skating, 163
 Yupik words for sea ice, xvii
Ingrams, Miss, 99–100
Inside Climate News, 245
Inside Science, 202
International Packaged Ice Association
  (IPIA), 78–79, 83
International Skating Union (ISU), 169–170
Inuit, xvii
Iñupiat, xvii, 253–254
Iroquois, 163
I See by My Outfit (Beagle), 80
Italian-style ices, 99
Ithaca Daily Journal, 106

J

J. C. Taylor & Son, *213*
Jackson, Augustus, 96–97
Jackson, Erin, 203
Jamaica, 13–14
Japan: methane hydrate extraction,
  256–258
Jefferson, Thomas, xvi, 94
Jensen, Larry, *199*
John Gorrie Museum, 34–35

Johnson, Richard, 185
Jones, Ada, 54
Journal of the Royal Society of Medicine, 231
Joy, Charles E., 152, 154
"Judy and the Iceman," 54
Julavits, Heidi, xx, 249
June, Charles, 204–205
Juvenile Sports and Pastimes, 185

K

Kansas City Star, 129
The Kentucky Housewife (Bryan), 114
Kerouac, Jack, 80
Kessler, John, 257
Kigali Amendment (2016), 247
King, Stephen, 201
Klineman, Jeffrey, 121
Knedlik, Omar, 80
Knickerbocker Ice Company, 58, 61, 66
Knickerbocker Trust Company, 67
Kold-Draft machines, 130, 136
Kubanda Cryotherapy, 227, 228
Kwan, Michelle, 177

L

lager, 25
Lake Michigan, 58
Lake St. Clair, 58
Lamond, Chris, 83–84
Laramy, Richard C., 81
Last Days of Knickerbocker Life in New York (Dayton), 98
Latini, Antonio, 94
Leadville, Colorado, 153–156
LeDuc, Timothy, 171–172
Lemp Brewery, 117
Le Petit, Joseph, 185
Letterman, David, 191
Letters on the Eastern States (William Tudor), 21
Liefferink, Rinse, 202
Lift Your Spirits (Williams and McMillian), 129, 131
Lincoln, Abraham, 33

Liquid Intelligence (Arnold), 129–130
Little Ice Age, xiv–xv, 59–60
Longfellow, Henry Wadsworth, 28
Louisiana Purchase Exposition (St. Louis World's Fair), 117–119, 120
Lynch, Mary, 205

M

Madison, Dolley, 94–95
Madison, James, 94–95
Madison Square Garden, 176, 189
Maggio, Maria, 83
magnetic refrigeration, 248
magnetocaloric material, 248
Mahapatra, Avipsa, 245–246, 247–248
manufactured ice
    during Civil War, 47
    economy and, 43, 47–48
    effects of, 34
    Gorrie's experiments with, 41, 42
    Gorrie's investment attempts, 43, 44
    Gorrie's marketing of, 44–45
    ice hockey and, 189
    for ice rinks, 173, 175–176
    in-house, 126
    machines for, 130, 136, 137, 138, 146
    marketing of, 44–45, 59
    New Orleans plant for, 127–128
    packaged, 78–79, 82
    public reaction to announcement of, 43
    smear campaign by Tudor, 45–46
    sold in fast-food restaurants, 79
    sports and, 49
    at St. Louis World's Fair, 118–119
    weather pattern changes and, 59–60
    during World War I, 69
    during World War II, 77–78
marketing
    cars, 81–82
    in Cuba, 19–20
    effectiveness of, xviii
    Gorrie's manufactured ice, 44–45
    ice cream, 103–104
    by manufactured-ice companies, 59
    of natural ice, 59, 60

marketing (*cont.*)
  in New Orleans, 23
  packaged ice, 78–79
  photography and, 120
  of refrigerators, 238–239, 241
Martinique, 11, 12–13, 16, 17
Mathisen, Oscar, 205
"Matilda Toots, or You Should Have Seen
  Her Boots" (De Marsan), 167
Mattus, Reuben, 107
Maude (television program), 241–242
"Maybellene" (Berry), 80
McAllister, Marvin, 99
McLean, Bobby, 205
McLeod, Chris, 137
McMillian, Chris, 129, 131
mechanical ice. *See* manufactured ice
medicine
  during Civil War, 217–220
  cryosurgery, 222, 226–231
  economics of, 229–230
  hypothermia experiments, 220–226
  ice as therapy, 216–217, 219–220
  ice for vaccine storage, 216
  temperatures and growth of cancerous
    tumors, 220–221
"Meet Me in St. Louis" (Sterling and
  Mills), 118
methane hydrate, 255–259
Middle East, sherbets, 93–94
Milk and Honey, 126–127
Miller, Christopher, 55
Miller Brewing Company, 26, 132
Mills, Kerry, 118
Minnesota Pure & Clear, 137
Mixology Ice, 137
Mobile Press-Register, 121
The Modern Steward (Latini), 94
Monahan, Kaspar, 179
Monticello, xvi, 94
Montreal, 149–151
Montreal Gazette, 186
Moore, Douglas, 106
Moore, Thomas, 24–25
Morgan, J. P., 67
Morse, Charles W., 61, 63–64, 66–68

Moss, Robert, 120–121, 122
motels, 82
Mount Vernon, 9, 91–92
Moya, Xavier, 249
"Mr. T." (Tureaud, Laurence), 193–194
Muller-McLave, Elsie, 205
Murphy, Charles F., 65

N

Nation, Carrie, 128–129
National Energy Conservation Policy Act
  (1978), 243
National Frozen Food Locker
  Association, 241
National Geographic, 253, 254, 257
National Hockey League, 189
National Housing Act (1934), 240
National Museum of American History
  (Washington, D.C.), 50
Native Americans. *See* Indigenous
  communities
natural ice
  climate change and, 236
  harvesting locations and health hazards
    of, 56, 57–59
  inventory tracking, 56–57
  marketing of, 59, 60
  as status marker, xvii
  Walden Pond, 235, 236
  as "white gold" in South, 40
  during World War I, 69
Natural Ice Association of America, 60
natural refrigerants, 247–248
Nestea, 122
Netherlands, 203
Neumann, Curt, 177–178
Névé Ice, 137
Newburgh, New York, 204–205
New Orleans
  gin fizz and, 129
  icehouses in, 127
  ice manufacturing plant in, 127–128
  during Prohibition, 131–132
  Frederic Tudor in, 21–24, 127
New Orleans Item-Tribune, 129

New York City
annual ice consumption in (1890), 61
deaths due to ice price hike, 62–63
Hippodrome, 174–176
Hippodrome Theatre, 176–177
ice deliveries, 55–56
iced tea in, 119
ice monopoly and politicians, 64–65
ice skating in Central Park, 165
pleasure gardens in, 97–98
safety of ice, 58
New-York Commercial Advertiser, 116
New-York Daily Globe, 46
New Yorker, xx, 239, 249
New York Evening Post, 65
New York Herald, 62, 101–102
New York Journal, 65
New York Times, 64, 65, 66, 67, 125–126, 166, 190–191
New York World, 62
Nutmeg Curling Club, 192, 197

O

Oelschlägel, Charlotte, 170, 177–178
Ogle, Maureen, 132
Old Absinthe House, 22
Olympic Games
curling at, 193–194
figure skating at, 167, 171, 172
speed skating at, 203, 205, 206–207
Zambonis used, 190
Olympic Oval (Utah), 206–209
O'Neill, Eugene, 54
One in a Million (movie), 178
"On the Prevention of Malarial Diseases" (Gorrie), 43–44
On the Road (Kerouac), 80
"open winters," 60
Otis, Harrison Gray, 11
Owen, Jake, 114

P

P. Diana and Sons Ice House, 84–85
P. T. Barnum's American Museum, 174

Pabst Brewing Company, 26, 132
packaged ice, 78–79, 82
Packaged Ice Quality Control Standards, 83
Panic of 1893, 184
Panic of 1907, 66–67
Parker Brothers, 191
Parkinson, Eleanor, 96, 102
Pattison, Mary, 238
Payne, Charles, 204
Peanuts (comic strip), 191
penny licks, 101
Persians, xvii
Petraske, Sasha, 126
Philadelphia
confectionery shops in, 96–97
ice skating on Schuylkill River, 164–165
safety of ice, 57–59
speed skating in, 204
photography, 120
Pike and North Lake Company, 26–27
Pittsburgh Commercial Gazette, 188
Pittsburgh Post, 188
Pittsburgh Press, 179, 188
plastic crystals, 249
plastic waste, 253
Platt, Chester, 106
pleasure gardens, 97–100
Ponzi, Charles, 67
Popsicles, 104
portable ice chests/coolers, *71*
invention of, 81
magnetic wine, 248
road trips and, 73, 81
Porter, Cole, 54
Powell, Marilyn, 93
Priessnitz, Vincenz, 216–217
Primack, Richard B., 236
Principles of Domestic Engineering (Pattison), 238
Prohibition, 105, 106, 131–132
Project Drawdown, 243
Pure Food and Drug Act (1906), 59
Pyke, Geoffrey, 157

Q

Quinzio, Jeri, 93

R

racial segregation
  of confectionery shops, 96–97
  of pleasure gardens, 98–99
  of soda fountains, 106
Ramos, Henry Charles "Carl," 128–129,
  130–131
Randolph, Mary, 95
Reagan, Ronald, 243–245
Rebel without a Cause (movie), 80
Reed, John Shelton, 122
Reed, N. B., 119–120
Rees, Jonathan, 54, 239
Refrigeration Nation (Rees), 239
refrigerators, *233*
  adoption of household, 236, 237, 238–
    239, 240, 241
  cooling operation of, 248–249
  DOMELRE, 238
  energy efficiency of, 236–237, 239–240,
    242, 244–248
  iceboxes and, 76–77, 237, 240
  ice-cream industry and, 107
  icemen and, 78
  importance of, in South, 119
  marketing of, 241
  mass production of, 76
  national standards for, 243–245
  natural refrigerants and, 247–248
  prevalence of, 81
  reliability of, 239
Reid, Robert Raymond, 40
Reid, Rosalie, 40
religion and manufactured ice, 42, 43, 44
Rickard, Tex, 189
road trips, 73, 80–82
Roaring Twenties, 74
Roberts, E. H., 77
"Rocket 88" (Brenston), 80
Rodden, Mrs. Pleas, *233*
Roman Hippodrome, 174–175
Roosevelt, Theodore, 63, 65
Rorer, Sarah Tyson, 116
Rosan, Monsieur (Florida's French
  consul), 45
Royal Ice Cream Parlor (Durham, North
  Carolina), 106
rum, 133
Ruppel, Carolyn, 255–256, 257–258, 259
Rural Electrification Act (1936), 77, 240
Russell, Joan, 205
Russell, William Howard, 166
Russia, 149

S

San Francisco, 135
Saturday Evening Post, 54, 115–116
Savage, James, 12–14, 18
"savior" ice (methane hydrate), 255–259
Sazerac, 24
Schenley Park Amusement Company,
  187–188
Schulz, Charles, 191
Schuylkill River, 57–58, 164–165
Scientific American, 43, 225, 237
Scott, John M., 105–106
Scudder, Peter, 99
seafood trade, 48
Seaton, Josephine, 95
Seinfeld, Jerry, 243
Semple, S. P., 114
Serious Eats, 120–121
7-Eleven stores, 79–80
Shadd, Sallie, 99
"shaker boys," 129
Shelton, Blake, 114
"shine sleighs," xv
"shinty," 184
Silver Skates Derbies, *199,* 206
Slate, 121
Slocombe, Humphry, 107
Slurpees, 80
smallpox, 218
Smarter Every Day (YouTube series), 195
Smith, Edwin, 216
Smith, Sydney, 133

Snapple, 122
soda fountains, 105–106
South
  annual yellow fever season in, 36–37
  ice as "white gold" in, 40
  importance of refrigerators in, 119
  manufactured ice and economy of, 47–48
  pre–Civil War sanitary conditions in, 37
  "sweet tea" and, 113–115, 120–122
Southern Cooking (Dull), 120
*Southern Food: At Home, on the Road, in
    History* (Egerton), 121
Southland Ice Company, 74–75, 77–78
speed skating
  early, 203–206
  at Olympics, 203, 205, 206–207
  Silver Skates Derbies, *199,* 206
sports, 49. *See also specific sports*
St. Lawrence River, 58
St. Louis and beer, 116–119
St. Louis World's Fair. *See* Louisiana
    Purchase Exhibition
St. Paul Winter Carnival, *141,* 151–153,
    156–157
Staff, Arthur, 206
Stedman, Charles, 117
Steel Magnolias (movie), 121
Steinbeck, John, 80
Sterling, Andrew, 118
"storming of the castle" tradition, 152
Stratton, Charles, 44
Strauss, Johann "the Waltz King," 175
Strong, William L., 62
Stubbs, Naomi J., 98
sundaes, 105–106
Surtees, Bailey, 227–230
swamp gas, 38–39
Swanson's television dinners, 241–242
"sweet tea," 113–115, 120–122
Syers, Madge, 170

T

Taft, William Howard, 67, 68
Taylor, Joe Gray, 119
Tebbutt, Charles Goodman, 184

television dinners, 241–242
temperance movement and iced tea, 115–116
temperature and humans, 39–40, 220–221
therapeutic hypothermia, 220–226
Thin Ice (movie), 178
Thomas, Jerry, 134–136
Thompson, Joe C., Jr., 74, 75–76, 79
Thoreau, Henry David, 235
The Three Stooges (film), 55
Tom Anderson's Saloon, 131
"Too Darn Hot" (Porter), 54
Tote'm Stores, 75–76, 79
Travels with Charley (Steinbeck), 80
Trump, Donald, 246
Truskolaski, Shane, 207–209
Tudor, Delia Jarvis, 5
Tudor, Euphemia, 28–29
Tudor, Frederic
  basic facts about, 5, 6, 27
  Cuba and, 6–8, 10–11, 18, 19–21
  death of John Henry, 9
  debt from ice business failure, 18
  Gorrie's manufactured ice and,
      45–46
  harvesting of first yield of ice, 15–16
  Hopkinson and, 27–28
  as "Ice King," 21
  ice-tool development, 24
  marriage, 28–29
  Martinique, 11–12, 16–17
  plan to ship ice to tropics, 10–15
  pleasure garden of, 98
Tudor, Frederic, Jr., 29
Tudor, John Henry, 6–9
Tudor, William, 12, 18, 21
Tudor, William, Sr., 5, 6
Tudor, Harry, 21–23
Tureaud, Laurence "Mr. T.," 193–194
Turner, William, 12–14
TV dinners, 241–242
Twain, Mark, 59

U

Unilever, 108
United States Shipping Company, 68

U.S. Figure Skating Association, 167
Utqiaġvik (formerly Barrow), Alaska, 253–254

V

Vaccaro, Pamela J., 119–120
vaccine storage, 216
Van Wyck, Augustus, 65
Van Wyck, Robert Anderson, 64, 65
Varon, Joseph, 225–226
Vauxhall Garden (New York City), 97
The Virginia House-Wife (Randolph), 95
Volstead Act, 131

W

Walden (Thoreau), 235
Walden Warming (Primack), 236
Washington, George, 9, 91–93
Washington, Martha, 9, 93
Washington Evening Star, 54, 60
water pollution, 57–59
Weather Up, 125–126, 127, 137
Weatherup, Kathryn, 125–126, 127
"We Gonna Move to the Outskirts of Town" (Weldon), 70
Weiss, Laura B., 105
Weiss, Pierre, 248
Weldon, Casey Bill, 70
Wesson, Wiley, 77, 78
Whirlpool Corporation, 247, 248
White, Robert, 223–225
"white gold," 40
Willard, Orsamus, 134
Williams, Elizabeth M., 129, 131
Williams, Virginia, 106
Wilson, Aaron, 204–205
Wilson, Charles Kemmons, Jr., 82
Wilson, Dorothy, 82

winter carnivals
    Montreal, 149–151
    St. Paul, *141*, 151–153, 156–157
"Winter Garden Glide," 187
Wisconsin Lakes Ice and Cartage Company, 26–27
Wolley, Charles, 165–166
women. *See also specific individuals*
    figure skating and, 165–167, 170, 177–178
    ice delivered by, *ii*
    ice skating and, 165–167, 169–172
    as salespeople for ice companies, 76
    at soda fountains, 106
    speed skating and, 203, 205, 206
Wondrich, David, 132, 133
Wood, Tim, 171
World Figure Skating Hall of Fame, 169
World War I, 68, 69
World War II, 77–78, 157
Wyeth, Nathaniel Jarvis, 24
Wynne, W. A., 64

Y

yellow fever
    in Apalachicola, Florida, 38
    causes of, 38–39
    effect on southern life, 36–37
    mortality, 36, 40
    other names for, 36
    treatment for, 37, 38
Yuengling, 132
Yupik, xvii
Yupik words for sea ice, xvii

Z

Zamboni, Frank, 190–191
Zambonis, 190–191, 194, 208–209